Pelican Books
Nuclear Power

Walter C. Patterson was born in Canada in 1936. He was educated in Manitoba, graduating in nuclear physics from the University of Manitoba. In 1960 he came to Britain to teach and write and became involved in environmental work in the late 1960s. He was editor of a small magazine, *Your Environment*, from 1970 to 1973.

In 1972 he attended the United Nations Stockholm Conference, as a member of the staff of *Eco*, the first of a regular series of independent conference newspapers. In the same year he joined the staff of Friends of the Earth Ltd, in London, as an energy specialist and he has been there ever since. A regular contributor to *New Scientist*, *Christian Science Monitor*, *Energy Policy*, *Environment* and other publications, he is also a consultant to radio and television broadcasting, particularly with regard to nuclear energy, and was one of the contributors to *The Shetland Way of Oil* (1976).

Walt Patterson is married to an Englishwoman and they have two daughters. Among his other interests are music, including Josquin and jazz, 'travelling, the slower the better; learning languages to the level of intelligible mediocrity; growing vegetables; brewing my own beer; consuming the last two items'.

NUCLEAR POWER

Walter C. Patterson

Penguin Books

Penguin Books Ltd,
Harmondsworth, Middlesex, England
Penguin Books, 625 Madison Avenue,
New York, New York 10022, U.S.A.
Penguin Books Australia Ltd, Ringwood,
Victoria, Australia
Penguin Books Canada Ltd, 2801 John Street,
Markham, Ontario, Canada L3R 1B4
Penguin Books (N.Z.) Ltd, 182–190 Wairau Road,
Auckland 10, New Zealand

First published 1976
Reprinted 1976, 1977 (twice), 1978

Copyright © Walter C. Patterson, 1976
All rights reserved

Made and printed in Great Britain by
Cox & Wyman Ltd, London, Reading and Fakenham
Set in Monotype Imprint

Except in the United States of America,
this book is sold subject to the condition
that it shall not, by way of trade or otherwise,
be lent, re-sold, hired out, or otherwise circulated
without the publisher's prior consent in any form of
binding or cover other than that in which it is
published and without a similar condition
including this condition being imposed
on the subsequent purchaser

To my parents, who didn't worry
when I dropped nuclear physics;
and to Cleone, who didn't worry
when I picked it up again

Contents

List of Figures   9
List of Tables   9
Preface   11
Introduction: The Nuclear Predicament   15

Part 1 The World of Nuclear Fission

1 What is a Reactor?   23
2 Reactor Types   42
3 The Nuclear Fuel Cycle   87

Part 2 The World and Nuclear Fission

4 Beginnings   117
5 Out of the Background   132
6 Reactors Off and Running   158
7 The Charge of the Light Brigade   189
8 Counting the Costs and Costing the Counts   218
9 Plutonium at Large   234
10 The Nuclear Horizon   257

Appendix A: Nuclear Jargon   269
Appendix B: Ionizing Radiation and Life   280
Appendix C: Bibliography: A Nuclear
  Bookshelf   286
Appendix D: Nuclear Organizations Pro and
  Con   297
Index   299

List of Figures

1 Nuclear power station   40
2 Magnox reactor   51
3 Advanced gas-cooled reactor (AGR)   58
4 High-temperature gas-cooled reactor (HTGR)   60
5 Pressurized water reactor (PWR)   64
6 Boiling water reactor (BWR)   68
7 CANDU reactor   71
8 Steam generating heavy water reactor (SGHWR)   74
9 Fast breeder reactor (FBR)   78
10 The nuclear fuel cycle   88
11 Atmospheric and underground explosions 1945–73
                                                            136–8
12 Power reactors in operation   259
13 Installed nuclear power capacity   260

List of Tables

1 Typical reactors   80
2 Nuclear explosions 1945–73   140
3 The growth of nuclear power   258

Preface

Nuclear reactors are fascinating. They are the heart of the technology which may shape the world of our near future, or may obliterate it. Born amid the tightest of military security during the Second World War, nuclear reactors have always impressed the layman as esoteric, fantastic entities, beyond ordinary comprehension. Such an impression is unwarranted. As we balance on the threshold of total commitment to a nuclear pathway it is vitally important that nuclear policy be based on broad public understanding – of nuclear technology, of its applications, and of its implications.

Nuclear physics and nuclear engineering are to be sure specialized subjects, dealing in phenomena which sometimes seem almost Carrollian in their unexpectedness. But the essential features of nuclear reactors have changed very little in the three decades of their existence; what has changed is their size, and their context. The present book is an attempt to describe the reactors themselves – their construction, behaviour, maintenance, offspring and relatives – and describe too the effect they have had, and are having, on the world in which we live. Of necessity the description is one man's view of an ever more controversial nexus of issues. It will strike some as unduly pessimistic about nuclear prospects, and others as entirely too absorbed in a subject from which they instinctively recoil. It is therefore appropriate to begin with a warning: in nuclear matters never take any one viewpoint as gospel, including this one. For those who wish to pursue questions further I have included in Appendix C (pp. 286–96) a lengthy list of additional sources, annotated to indicate – once again – one man's view of their virtues and shortcomings.

In the course of my own involvement with nuclear matters I have been fortunate to encounter many other viewpoints against which to measure my own, some in print and some in person. The staff of the United Kingdom Atomic Energy Authority have been unfailingly helpful to me in spite of our frequently diverging opinions. I should particularly like to thank Ron Truscott and his colleagues of the Public Relations section, and Mrs Lorna Arnold of the Archives, who made available to me advance copies of Professor Margaret Gowing's superb official history *Independence and Deterrence* at a crucial stage in my own research.

The now sub-divided United States Atomic Energy Commission has supplied me with many useful documents, as have the International Atomic Energy Agency and the OECD Nuclear Energy Agency, the last through the good offices of Bruce Adkins, to whom again my thanks. I have made extensive use of the works of Dr Samuel Glasstone, the late Dr Theos Thompson and Dr J. G. Beckerley, the late Dr Kenneth Jay, my friend and colleague Sheldon Novick, Dr J. E. Coggle, Dr Tom Cochran, Dr John Gofman, Dr Arthur Tamplin, Dr John Holdren, Richard Lewis, John McPhee, Norman Moss, Dr Peter Metzger, Roger Rapoport, Ted Taylor and Mason Willrich, among many; to all of them my thanks. The publications of the Stockholm International Peace Research Institute, the Pugwash Conferences, the International Commission on Radiological Protection, the National Radiological Protection Board and the Union of Concerned Scientists, and the pages of the *Bulletin of the Atomic Scientists*, *Science*, *Environment*, *Energy Policy* and *Nuclear Engineering International* have yielded much valuable material, as has the *Weekly Energy Report*, to whose editor and publisher Llewellyn King my thanks.

Many of my journalist colleagues have for many months carried on with me an unending and fruitful colloquy, not only about nuclear affairs but about their context in energy and social policy overall. I expect they will let me know what they think of my effort, in similarly blunt language: in anticipation my thanks to the lot of them, especially to the staff of the *New Scientist*, who encounter my opinions sooner and more often than most.

Even closer to home come my colleagues in Friends of the Earth International, without whose active participation this book neither would nor could have been written. Brice Lalonde and Pierre Samuel of Les Amis de la Terre in France, Lennart Daléus of Jordens Vänner in Sweden, Brian Hurley of FOE Ireland, Holger Strohm of Die Freunde der Erde in West Germany, Kitty Pegels of Vereniging Milieudefensie in the Netherlands, Jim Harding of FOE Inc. in the USA, and many other Friends have provided an international link-up of growing strength. In Britain Dr Peter Chapman of the Open University and Gerald Leach of the International Institute for Environment and Development have contributed provocative insights and stimulating debate. My Friends of the FOE Ltd office in London have put up with my absence from the team for some weeks without complaint, carrying for me my share of the load. Finally, and most particularly, my warmest thanks to Dr John Price and to Amory Lovins, with whom it is an exhausting pleasure to work, and without whom I might as well stick to cultivating my broccoli.

I am grateful to Peter Wright of Penguin Books, for the opportunity to amplify fourfold an earlier version of this text. On the home front my thanks and best wishes to Mrs Sue Hunter, who has toiled nobly to turn a tangled typescript into intelligible copy. Lastly, to my beloved wife Cleone, who put up with three months of me lying obsessively awake at four in the morning: I promise that I shall never again write about terrorists with nuclear weapons the day before Christmas Eve. I hope.

Walt Patterson
31 December 1974

Introduction:
The Nuclear Predicament

For three decades the world has been learning to live with nuclear energy. The learning process has been exciting, frustrating, and sometimes frightening; it is far from over. Indeed it may be just beginning. We have learned a great deal about how to release nuclear energy; how to control it; and how to make use of it. We have even learned to take it for granted. But we have not yet learned to live with it. Nuclear energy in all its aspects is already shaping the world. The future of our globe will depend to a startling extent on what we know about nuclear energy, and what we do about it. The crucial decisions will not wait another three decades.

Concentrated high-quality energy has become a staple commodity in our industrial society. The most concentrated energy available is nuclear energy, made accessible by nuclear reactors. The energy contained in 1 kilogram of uranium, if it were all to be released in a nuclear reactor, would be equivalent to that produced by burning some 3000 tonnes of coal. It is not, of course, quite that simple; the possibly apocryphal British workman who filched some uranium reactor fuel and tried to burn it in his grate was disappointed. But there is no doubt that the world's reserves of uranium represent a staggering store of energy. If suitable reactors and other facilities are provided, it becomes possible to exploit uranium, which would otherwise be virtually useless. The same is true of the even more plentiful metal thorium.

These constructive possibilities were identified very early in the development of nuclear energy, even as plans were taking shape to release nuclear energy explosively. The awesome destructive power of nuclear weapons dominated the scene for the first post-war decade. But by the mid 1950s scientists and engineers were well on the way to harnessing this power for peaceful purposes.

The prospects looked brighter and brighter. There had, to be sure, been a surge of euphoric predictions in the aftermath of the two nuclear explosions over Japan which ended the Second World War. So-called 'atomic power' would run a car on an engine the size of a fist; we would soon live in houses heated by uranium; 'atom-powered' aircraft would be able to remain aloft indefinitely; 'atom-powered' rockets would enable us to cross the ocean in three minutes – and so on. But the people who really understood the implications of nuclear energy were much more realistic. They chose applications whose development seemed fairly straightforward; and their efforts bore fruit.

A nuclear reactor releases nuclear energy in the form of heat; the heat is used to generate steam, and the steam to generate electricity – with conventional electrical equipment. Since the mid 1950s nuclear generation of electricity has become a full-fledged technology, now on the verge of a vast world-wide expansion. From the outset it seemed that the nuclear approach to electricity generation would have certain advantages and disadvantages, in comparison with conventional generating stations which raise steam by burning coal, oil or gas. Fossil-fuelled power stations are less expensive to build than comparable nuclear power stations. On the other hand, it was expected that the running cost of a nuclear power station would be considerably less than that of running a fossil-fuelled station. Some early publicity went so far as to declare that nuclear electricity would be too cheap to meter. But as usual those on the inside made no such claim. Instead they calculated the total cost of a unit of electricity generated by a fossil-fuel or by a nuclear station, taking into account both capital and running costs. Estimates differed, but there was every likelihood that a unit of nuclear electricity would cost only about one-fifth as much as a unit of fossil-fuel electricity. On this basis nuclear power stations looked an excellent investment.

In the ensuing years, the bases for these economic calculations varied. For a time the cost of oil remained low, while that of coal increased; some nuclear costs also increased, and the balance remained uncertain. By the late 1960s mounting public concern for the environment was drawing attention to the problems arising from large-scale use of fossil fuel: health hazards from underground coal-mining, ecological damage from surface mining, marine pollution from transport of oil, and air pollution from the burning of coal and oil. By contrast nuclear power stations seemed environmentally inoffensive.

In the early 1970s the upward surge of oil prices, and increasingly uneasy labour relations in the coal-fields, added to the comparative economic attractions of nuclear power. The gradual and tentative industrial commitment to nuclear power began to accelerate dramatically (see Figure 12, p. 259 and Table 3, p. 258). So did the nuclear component of total electricity output (see Figure 13, p. 260 and Table 3, p. 258).

Governments wanted to lessen their dependence on the petroleum-exporting countries; electrical supply systems wanted to lessen their dependence on coal, especially because of their vulnerability to recalcitrant unions. Nuclear electricity generation seemed the obvious alternative. In the longer term, it was argued, coal and oil would become irreplaceable raw materials for the chemical industry, and should be reserved for these uses, while nuclear energy was used for electricity. Electricity, it was further argued, is a premium form of energy, versatile, high-grade and clean at the point of use. It ought accordingly to be an ever-larger proportion of total energy used. Since up to that time nuclear sources could most readily be used to provide electricity it all seemed to fit together very neatly. World energy use would continue to rise rapidly; so would energy use per person, as more and more people shared in the benefits of modern technology. One authoritative view foresaw a world in which world-wide energy consumption *per capita* would be twice that of present-day Americans – this energy would be provided by some 4000 clusters of nuclear power stations, each cluster containing enough reactors to produce five times the output of today's largest power stations. For such a high-energy future the role of nuclear energy would be crucial. Only by the most vigorous possible growth of nuclear capability could mankind's energy requirements be met.

Such an argument was and is persuasive. It is not, however, unanswerable; and while some voices were calling for more and larger reactors as fast as possible, other voices were asking other questions, some of which are also difficult to answer.

The earliest questioning derived from lingering public fear and distrust of nuclear energy, because of its first appearance as the most devastating weapon ever used. Gradually, certain specific issues crystallized out of the general unease. The world has somehow got used to the overwhelming destructive power stored in the nuclear arsenals of the USA, the USSR, the UK, France and China; few would hesitate to identify these arsenals as the most terrible threat to the future of life on earth. But apart from these explicit military aspects of nuclear energy, several other aspects also give rise to concern. We shall examine these in some detail in the coming chapters. It must here suffice to mention them briefly, as issues which will recur repeatedly in later discussion.

Nuclear reactors and other nuclear facilities produce and contain enormous quantities of material which is 'radioactive' (pp. 25–8). Some radioactive materials are very dangerous to living things, and many remain so for unimaginably long times. These materials must on no account be allowed to escape in quantity from nuclear facilities. Such facilities release small amounts of radioactivity to their surroundings during normal operations. One area of bitter controversy concerns the standards and controls applied to these releases. Some critics with impressive credentials consider present standards far too lax, especially in view of the anticipated upsurge in the number and size of nuclear installations. Another major concern is operating safety, not only of the various designs of reactor themselves but also of their support facilities including transport systems. It has of late become unpleasantly clear that such safety must take into account the possibility not only of accidents but also of sabotage. A protracted expert disagreement about safety has recently overtaken the most popular designs of reactor, and continues unabated. Other designs have not thus far been subjected to such intense independent scrutiny.

One category of radioactive material arising from nuclear activities requires particular mention. This is the 'high-level' waste left after chemical processing of used reactor fuel (see pp. 108–14). High-level waste contains large amounts of substances which are dangerously radioactive, and will remain so for hundreds of years. What to do with these wastes is a question as yet unanswered. Provisional answers have been proposed, and interim management is said to be adequate, but in the long term the question becomes one not of technology but of ethics. Should we create these dangerous substances in ever-increasing quantities, to leave them to our remote descendants?

Ethics aside, it has become gradually apparent that considerations of safety affect the overall cost of nuclear power. Just as coal-mining must take account of the cost of health measures, land reclamation and pollution control, the use of nuclear power must allow for costs of extra safety measures and related provisions. The optimistic cost comparisons, originating in the early 1950s, which favoured nuclear over fossil fuels, are still, to be sure, impressive – but they no longer look unarguable, as we shall describe.

Simple expenditure, however massive, seems unlikely to provide the requisite guarantees for one aspect which might narrowly be related to safety. As the world economy comes to rely more and more on nuclear reactors as a source of power, so the traffic in 'fissile' materials increases – materials which can be made into nuclear weapons. Authoritative studies have shown that present provision for the security of these materials is frankly perfunctory. The prospect of nuclear weapons in the hands of unstable governments, terrorist organizations or deranged fanatics is not one calculated to encourage a rosy view of the global future. Current suggestions for dealing with this danger include special government nuclear police forces, relentless official probing of personal histories of nuclear employees, monolithic central administration of public policy, and other proposals which come uncomfortably close to outlining a virtually totalitarian social structure.

It becomes apparent, when considering these intractable

problems, that the decisions we make about nuclear energy will determine in large measure the kind of world our grandchildren will inherit. The issues this technology has created constitute a remarkable microcosm of the present predicament of our planet. The nuclear predicament raises a host of social, political and even ethical problems, many of them with long-term implications beyond any foreseeable horizon. Clearly such issues demand the fullest public consideration, the widest possible participation in the crucial decisions to come.

Public participation in nuclear decision-making has hitherto been either tentative or desperate, largely because the issues seem to be cloaked in the most esoteric of scientific obscurity. But the veil of mystery surrounding nuclear matters has always been primarily one of military secrecy, not of intellectual in-accessibility. In the next three chapters we shall describe why and how nuclear reactors work, and the other services they re-quire. If you are a nuclear engineer you can skip these chapters. If not, you should read them carefully, as they make it easier for you to determine whether nuclear engineers are talking sense.

Part One

The World of Nuclear Fission

1. What is a Reactor?

Atom and Nucleus

If you take a pair of metal hemispheres and slam them together very fast face to face, one of two things may happen. You may get a loud clunk. Or you, the hemispheres and everything else in the vicinity may be almost instantly vaporized in a burst of incredible heat. If the latter happens, you can be sure that the metal was a particular kind of uranium, not that the confirmation will do you much good.

What has vaporized you is raw energy, released from the innermost structure of the uranium. The energy in the interior of uranium was revealed to the world on 6 August 1945, in the sky above Hiroshima, Japan. Never has a source of energy made a more horrifying debut. Yet, paradoxically, the most over-powering energy man has learned to release comes from the tiniest reservoir he has yet learned to tap: the nucleus of an atom.

What is an 'atom'? And what is its 'nucleus'? Suppose you take a lump of lead, and cut it into smaller and smaller pieces. When the pieces are so small that your knife is too clumsy, switch to an imaginary knife and keep cutting. Ultimately, the pieces will get so small that if you cut any more you will not get two pieces of lead: the next cut will change the identity of what you are cutting. The smallest piece which is still lead is called an atom of lead.

The word 'atom' means 'indivisible'. You cannot divide an atom of lead and still get lead. But you can divide the atom and get smaller pieces which are no longer lead. If you start to dismantle an atom, the first part you get off is called an 'electron'. Until this stage you have been able to cut without encountering any conspicuous electrical effects; but the electron is negatively charged, and the remainder of the atom is left positively charged. The parts of the atom have become 'ions':

a negative ion (the electron) and a positive ion (the remainder of the atom). Each time you remove another electron you leave still more positive charge on the remainder. It becomes doubly 'ionized', triply 'ionized', and so on. Since negative and positive charges attract each other, it is harder and harder to pry off successive electrons.

Suppose, however, that you manage to remove *all* the electrons. (For most atoms this is in practice very difficult.) What you have left is the innermost heart of the atom: the nucleus. This is where all the positive charges are. Furthermore, you would now find an enormously increased difficulty in cutting any more. Surprisingly enough, although the nucleus contains only positive charges (which *repel* each other), its constituent pieces cling together with a loyalty which makes the outer electrons look frankly promiscuous.

There used to be much popular talk of 'splitting the atom', but the problem was rather one of splitting the *nucleus* of the atom. 'Atomic' bombs should have been called 'nuclear' bombs: for the shattering energy they released came from the rupturing of nuclei (one nucleus, two or more nuclei).

Uranium

What makes uranium so dramatically different from other substances? To appreciate its unique characteristics we must first consider some basic nuclear physics: that is, what nuclei consist of and how they behave. An atom is made of electrons around a nucleus; a nucleus in turn is made up of 'protons' and 'neutrons'. A proton has a positive charge; a neutron has no electrical charge and is 'neutral'. At first it seems difficult to understand how a nucleus stays together at all. The positive charges of the protons ought to push them violently apart. But within the compact volume of the nucleus a new kind of force comes into effect: an immensely powerful short-range attractive force acting equally between protons and neutrons – which, from this point of view, are all 'nucleons'. The short-range nuclear force holds them together, against the repulsive effect of the protons' positive charges. In this way the neutrons act as 'nuclear cement'.

However, in a nucleus which contains 92 protons – that is, a nucleus of uranium – the repulsive force among the protons is on the verge of overcoming the nuclear force. If there are as many as 146 neutrons also present, the nucleus can remain intact – barely. This form of uranium, containing in all 238 nucleons, is called uranium-238 or $^{238}_{92}U$. For reasons that need not concern us here, involving the grouping and compatibility of nucleons, the next most probable arrangement is a uranium nucleus containing three fewer neutrons: uranium-235, $^{235}_{92}U$. Atoms with these lighter nuclei make up about 0.7 per cent of naturally-occurring uranium. (If nuclei have the same number of protons, they are nuclei of the same chemical 'element': thus, every nucleus with 92 protons is the nucleus of an atom of uranium. Atoms whose nuclei have the same number of protons but different numbers of neutrons are called 'isotopes' of the element: for instance, uranium-238 and uranium-235 are isotopes of uranium.) The uranium-235 nucleus has a property unique among all the more than 200 types of nuclei found in nature in significant quantity before 1942. The uranium-235 nucleus is already under near-disruptive internal stress; a stray neutron blundering into it can rupture it completely.

Radioactivity Produces Radiation

When a stray neutron hits a uranium-235 nucleus, the result is a 'compound nucleus' of uranium-236. It is called a compound nucleus because it does not last long. The energy added by the neutron – even a 'slow' one – overcomes the precarious stability of the nucleus and, almost instantly, it flies apart. The rupture of a uranium-236 compound nucleus usually results in about two-fifths of the nucleus flying off in one direction and about three-fifths in the opposite direction, with perhaps two or three odd neutrons also shooting out. The flying fragments burst out with so much energy that a subsequent tally of masses reveals a shortage: some of the mass of the original nucleus has been converted into energy. This is the source of the enormous energies released in such nuclear events.

For example, one common subdivision results in one chunk of 38 protons and 52 neutrons, another of 54 protons and 89 neutrons, and 3 odd neutrons; making up, of course, 236 nucleons in all. The chunk containing 38 protons is a nucleus of strontium; since it contains in all 90 nucleons it is the notorious strontium-90. The chunk containing 54 protons is a nucleus of the inert gas xenon; since it contains in all 143 nucleons it is xenon-143.

Such a complete rupture of a nucleus is called a 'fission', by analogy with the biological term for the division of a growing cell. More precisely it is called 'nuclear fission'. When it is provoked by the impact of an additional neutron it is called 'induced fission'; such is the case with uranium-235 just described. Some very heavy nuclei are so unstable that they may rupture even without being struck by a neutron; such a rupture is called 'spontaneous fission'. Fission, whether induced or spontaneous, is the most violent kind of breakdown that a nucleus can experience. But there are others. A nucleus of uranium-238, for instance, while not being so near to rupture as its lighter relative, is still under severe stress: so much so that sooner or later it is likely to squirt out a lump made up of two protons and two neutrons. Since this makes a larger proportional reduction in the proton contingent than in the neutron contingent, the remaining nucleus, now containing only 90 protons and 144 neutrons, is slightly less stressed. (It is a nucleus of the metal thorium-234.) The lump or 'particle' squirted out is identical in every respect to an ordinary helium nucleus; but since it emerges with considerable velocity, and ploughs a furrow through whatever slows it down, it is given a special name: it is an 'alpha particle'. Most nuclei with at least 83 protons undergo a breakdown this violent; they are called 'alpha emitters'.

The balance between protons and neutrons in thorium-234,

while more satisfactory, is far from ideal. In effect, by emitting an alpha particle, the nucleus has overadjusted. This leads to a yet more delicate form of breakdown. Out of the nucleus containing 90 protons and 144 neutrons there suddenly squirts an electron. It is identical in every respect to the electrons outside the nucleus; but since it emerges with considerable velocity it too is given a special name: it is a 'beta particle'. The nucleus which remains now contains one more positive charge than it had. But since an electron is very much less massive than a nucleon, there are the same number of nucleons as before: 234. A neutron has apparently turned into a proton. The nucleus now contains 91 protons and 143 neutrons; it is a nucleus of protactinium-234. Like thorium-234, protactinium-234 is also a 'beta emitter'; when it emits a beta particle it becomes uranium-234, which is an alpha emitter. So, leapfrogging down by alternate alpha and beta emission, the nucleus alters itself until it has only 82 protons and 124 neutrons, and is at last stable: a nucleus of lead-206.

On the way, the nucleus regularly finds itself, after emitting an alpha or beta particle, still unduly agitated or 'excited'. To settle itself down it gives off a burst of energy in a form closely akin to ordinary light, but much more energetic, and invisible. This burst of energy is called a 'gamma ray'. It is identical in every respect to the well-known 'X-ray', except that an X-ray comes from the electron layers outside the nucleus, whereas a gamma ray comes from inside the nucleus.

Consider also the nucleus of strontium-90, one of the two large fragments formed by the induced fission of uranium-235 in the earlier example. The strontium-90 nucleus, a 'fission product', has a disproportionately large number of neutrons for its protons, coming as it does from a much heavier nucleus which requires more 'cement'. Accordingly, the strontium-90 nucleus is also a beta emitter. Sooner or later it squirts out a high-velocity electron – a beta particle – and one of its neutrons is replaced by a proton. It becomes a nucleus of yttrium-90, another beta emitter, which by the same process becomes a nucleus of zirconium-90, which is stable. Beta emissions from fission-product nuclei are often followed by one or more gamma rays.

There are thus four ways in which a nucleus can alter itself: fission, alpha emission, beta emission, and gamma emission. From a lump of material containing such unstable nuclei the emissions from these activities shoot out radially in all directions; the lump is said to be 'radioactive', and the emissions – neutrons, alpha and beta particles, and gamma rays – are called 'radiation'. A collection of nuclei which shoots out 37 000 000 000 such emissions per second is said to exhibit one 'curie' of radioactivity. (This is the radioactivity of one gram of radium, one of the first substances known to be radioactive, which was discovered by Marie Curie.)

In a radioactive substance it is impossible to tell whether a particular nucleus is on the point of radioactive breakdown, or 'decay'. Nonetheless, in a sufficiently large sample of any particular radioactive nuclear species, or 'radioisotope', a certain fraction of the nuclei always decay in a quite regular length of time. For instance, if you start with 1000 nuclei of strontium-90, 28 years later 500 will have decayed and you will have 500 left. After a further 28 years, 250 of the remaining 500 will have decayed, and you will have 250 left. And so on: however much you start with, 28 years later half will have decayed and only half will be left. Obviously the corresponding radioactivity will also have fallen by one-half. For strontium-90 the period of 28 years is called its 'half-life'. Each radioisotope has a half-life for each form of radioactivity it exhibits: in each case the half-life is the time during which half the nuclei in a sample decay, and the corresponding radioactivity falls to half the initial level. Half-lives of different radioisotopes range from fractions of millionths of a second to millions of years.

The Effects of Radiation

Unless radioactive decay takes place in a vacuum, the radiation emitted must pass through the surrounding substance. The consequences depend on the substance, on the type of radiation, on its energy and on its intensity. An alpha particle, made up of four nucleons with two positive charges, interacts vigorously with surrounding atoms, tearing off electrons and knocking nuclei out of place. In doing so, the alpha particle quickly gives up its own energy, travelling only a short distance but doing enormous damage along its path. Most alpha radiation is stopped within the thickness of a single sheet of paper. A beta particle, much less massive and with only one negative charge, disturbs and dislodges neighbouring electrons, but loses its energy less swiftly and therefore travels somewhat farther than an alpha particle. Most beta radiation is stopped within the thickness of a thin sheet of metal. A gamma ray, with no electrical charge, loses its energy much more gradually, and can travel a long distance, causing a relatively small amount of disturbance at any particular point on its path. A neutron, also without electrical charge, is likewise free to travel a long distance, and is slowed down mainly by direct collision with nuclei. Gamma or neutron radiation can penetrate more than a metre of concrete.

Dislodging an electron from an atom makes the atom an ion: so emissions from nuclei are 'ionizing radiation'. When ionizing radiation passes through a material it causes changes in the structure of the material – sometimes temporary, sometimes permanent, sometimes useful, sometimes harmful. The effects of ionizing radiation depend roughly on how much energy the radiation releases into a given amount of material – the more energy, the more disruption. The basic unit of radiation exposure is the 'roentgen', named after Wilhelm Roentgen, discoverer of X-rays.

The effects of ionizing radiation become particularly important if the radiation is passing through living matter; the delicate molecular arrangements of living matter can be easily upset by radiation. There are several units used to measure radiation effects on living matter. The most common are the 'radiation absorbed dose', or 'rad', and the 'roentgen equivalent man', or 'rem'; the latter allows for the greater severity of alpha or neutron damage for equivalent energy-delivery. For beta and gamma radiation one rad is about the same as one rem; for neutrons and alpha particles one rad may be up to twenty rem, depending on the energy of the particles.

The question of the biological effects of radiation is surrounded by controversy. But it is known that a dose of perhaps 400 rem of radiation over the whole body will kill half the adult human beings exposed to it; and very much smaller doses will produce cell damage that may lead to leukaemia and other kinds of cancer. Furthermore, radiation damage to the complex molecules in the reproductive cells which contain the hereditary information may produce mutant offspring. Even a single gamma ray can disrupt a gene; it may produce unforeseeable effects if this particular gene should be in a reproductive cell which subsequently helps to form a child.

A more detailed discussion of radiation biology is given in Appendix B (pp. 280–85). Suffice it to say here that the danger of radiation to living matter seems to increase in direct proportion to the amount of radiation exposure, beginning from the very lowest doses. There does not appear to be a threshold dose – that is, one below which damage does not occur. We are already subjected to continual radiation from the natural radioactive substances in our surroundings, and from cosmic rays. Any human activity which tends to add further sources of radiation to our surroundings must be potentially harmful. Just how harmful – and in return for what benefits – is still under debate; this book is intended to make one aspect of the debate more intelligible, whatever your criteria.

The Chain Reaction

In a lump of uranium there are always a few stray neutrons, produced either by spontaneous fission or by cosmic rays. Suppose that one of these stray neutrons induces a nucleus of uranium-235 to undergo fission. As well as the two fission products, there shoot outward perhaps two or three high-energy neutrons. (The chances are better than 99 to 1 that these neutrons will emerge virtually at the instant of fission: 'prompt' neutrons. But there is a slight chance that a neutron will not emerge until some seconds later: a 'delayed' neutron. As we shall see, delayed neutrons are of considerable importance.) There are three possibilities open to the high-energy neutrons from fission. A neutron may reach the surface of the material and escape. It may strike another nucleus and be absorbed without causing any immediate breakdown. Or – most importantly – it may strike another nucleus and, in turn, cause this nucleus to rupture. The chances of a neutron causing such an induced fission depend on the neutron's energy and on the nucleus it strikes. A fast neutron, fresh from an earlier fission, can cause a nucleus of uranium-238 to rupture; but such a fast neutron is not the most effective means of rupturing a nucleus of uranium-235. If a neutron ricochets among other nuclei, bouncing off each and giving up its energy bit by bit, it soon slows down until it is just jostling with the shared heat-energy of the rest of the material. It is then a 'thermal neutron'. Only a fast neutron can rupture a nucleus of uranium-238. But a thermal neutron, rather than a fast neutron, is much more likely to rupture a nucleus of uranium-235.

If, in a lump of uranium-235, one nucleus undergoes fission, the neutrons it releases may strike other nuclei, causing more fissions and releasing more neutrons. If there are enough uranium-235 nuclei sufficiently close together, the spreading disruption multiplies with astonishing rapidity: more and more neutrons, more and more ruptured nuclei, their fragments flying, more and more energy: a 'chain reaction'. If there is enough uranium-235, packed closely together for long enough, and if the chain reaction is out of control, the result is a nuclear explosion: an 'atomic bomb'. Slamming two suitable hemispheres of uranium-235 together at a very high velocity will indeed create a nuclear explosion; but there are other, much more efficient, techniques – and materials.

Needless to say, as soon as an appropriate arrangement of appropriate material became possible it was tried out – on 16

July 1945, at the top of a tall tower in the desert near Alamo-gordo, New Mexico: the world's first nuclear explosion, code-named Trinity. Within three weeks an atomic bomb made of uranium-235 devastated Hiroshima. But, seemingly, one 'doomsday weapon' was not enough, and uranium-235 was not the only nucleus that could be used. A neutron can penetrate a nucleus of uranium-238 without rupturing it. If this happens, the resulting neutron-heavy nucleus soon emits a beta particle, and then another, to become a nucleus of plutonium-239. Like uranium-235 – and only a few other isotopes, all at present very rare – plutonium-239 is 'fissile': that is, it can undergo a chain reaction of successive fissions, as the Trinity test demonstrated. On 9 August 1945 such a chain reaction obliterated Nagasaki.

The Nuclear Reactor

If chain reactions in uranium-235 and plutonium-239 could be used only in weapons, the situation would already be sufficiently complicated. But, more than two years before enough pure fissile material of either kind had been accumulated to make a weapon, it was found possible to control a chain reaction: to have it maintain itself without multiplying out of control. Indeed it was by this means that the plutonium was produced for the Alamogordo and Nagasaki bombs. The arrangement used to create and control a sustained nuclear chain reaction is called a 'nuclear reactor'.

The difference between an uncontrolled and a controlled chain reaction is profound. An arrangement of fissile nuclei which is to undergo an uncontrolled chain reaction – a nuclear explosion – must be sudden and final. An arrangement of fissile nuclei which is to sustain a continuing controlled chain reaction must be much more carefully organized. Curiously enough it takes many more nuclei – that is, much more material – to build a reactor than it takes to set off an explosion. This is of course partly because an explosion requires comparatively pure fissile material; in a reactor the fissile material is comparatively dilute, and there is accordingly much more material in all. But there must also be many more fissile nuclei themselves. The reason for this is the role played by the all-important neutrons.

If a chain reaction is to be self-sustaining it must keep itself supplied with neutrons. Consider the following typical sequence. A neutron plunges into a nucleus of uranium-235. The nucleus ruptures; as well as two fission-product nuclei it also shoots out three neutrons. One of these three goes out through the surface of the lump of uranium and is lost. Another is absorbed by a nucleus of uranium-238, which begins its two-stage change into plutonium-239 but does not rupture. This leaves one neutron. If this third neutron now plunges into another uranium-235 nucleus and ruptures it, the process can continue; otherwise the chain reaction is snuffed out.

At any instant, inside the lump of uranium, there must be the right number of neutrons of the proper energy to propagate the chain. In effect, for a sustained chain reaction, each neutron which is lost by causing a fission must be replaced by exactly one neutron which does likewise. The system then has a 'reproduction factor' of 1. When this condition is achieved the system is said to be 'critical', and the situation is called 'critical-ity'. (Note that 'criticality' is not here used to imply 'danger'.) If on average each neutron lost when it causes a fission is replaced by more than one which also causes fission, the reaction 'runs away'; the reproduction factor is greater than 1, and the system is 'divergent'. If on average each neutron so lost is replaced by fewer than one which causes fission, the reaction will stop; the reproduction factor is less than 1. This is why a piece of uranium below a certain minimum size cannot under normal conditions sustain a chain reaction: there is too much surface through which neutrons can leak out.

Moderators

The basic requirements for a continuing controlled chain reaction are therefore, first, a collection of fissile nuclei appropriately distributed in space; and, second, a self-replenishing supply of neutrons of just sufficient numbers and energy to keep the chain reaction going. In natural uranium, only 0.7 per cent of the nuclei are fissile uranium-235. These fissile nuclei, only seven out of every thousand, are not sufficiently close together to keep up a chain reaction; too many neutrons are absorbed by the heavier uranium-238 nuclei, without causing fission. To improve the prospects for a sustained chain reaction it is necessary either to increase the proportion of uranium-235 relative to uranium-238; or to slow down the neutrons to thermal energies, at which they are much more readily absorbed by uranium-235; or to do both.

As we shall see (pp. 91–5), increasing the proportion of fissile uranium-235 – so-called 'enrichment' of the uranium – is a complex and expensive process. But even a small increase, say from 0.7 per cent to 2 or 3 per cent, makes a marked difference, provided that the neutrons from fission are slowed down. This can be done by means of a material with light nuclei – a 'moderator'. A fast neutron striking a light nucleus in the moderator gives up a fraction of its energy, and after a few such collisions has slowed to thermal energy. The best moderators are the lightest nuclei: those of hydrogen. Ordinary water, containing two hydrogen atoms per molecule, is a satisfactory moderator. But ordinary hydrogen nuclei absorb neutrons. Better still is a rarer form of hydrogen nucleus: the proton-plus-neutron form called 'heavy hydrogen' or 'deuterium'. If two atoms of heavy hydrogen combine with an atom of oxygen, the result is a molecule of 'heavy water', or deuterium oxide (sometimes written D_2O), which is much the best moderator of fast neutrons.

One other substance is widely used as a moderator: carbon, in the form of graphite. A carbon nucleus – six protons and six neutrons – is much more massive than either form of hydrogen nucleus, and is therefore not such a good moderator. But graphite is less expensive than heavy water; furthermore it is a solid, which can be structurally useful in a reactor.

Reactor Design and Operation

To set up a nuclear reactor you proceed as follows. You take a good many pieces of material containing uranium-235 – usually uranium metal or oxide, natural or enriched: the 'fuel'. (You can also use plutonium-239, although this – as we shall see – involves some difficulties.) For a large reactor you need many tonnes of fuel, much more than enough to achieve criticality. One obvious reason for the extra fuel is to enable you to operate the reactor for some time before replacing the fuel. Other reasons will become clear in a moment.

You seal the pieces of fuel into casings called 'cladding', to support the fuel and to confine the fission products that will be produced. You position the assemblies of sealed fuel, called 'fuel elements', supporting them as necessary; remember that they may be very heavy indeed. You intersperse the fuel elements with moderator, to slow down the neutrons, and with neutron absorber, to enable you to control the chain reaction. You also include measuring instruments to tell you what is going on inside the reactor. You need to know, in particular, the temperature and the concentration of neutrons at various places inside the reactor.

You are now ready to start up your reactor. Before start-up, with all the absorbers in the interior of the reactor soaking up neutrons, the neutron density is so low it is difficult to measure, unless you intentionally include a separate source of neutrons as a sort of primer. A common form of absorber is a rod thrust through the interior of the assembly: a 'control rod'. Such a rod incorporates a material like boron, which absorbs neutrons like a sponge. The rod may be made, for instance, of boron steel. So long as enough control rods are in place no chain reaction is possible. To start up your reactor – to make a chain reaction possible – you begin withdrawing control rods.

The region of the reactor in which the reaction takes place is called the core. You withdraw control rods very slowly from the core, usually in short steps, in suitable symmetry to maintain a more or less uniform build-up of neutron density inside the reactor. In due course your reactor 'goes critical': a self-sustaining chain reaction is established, in which each neutron lost by causing a fission is replaced by exactly one neutron (either prompt or delayed) which does likewise. If the chain reaction could be sustained by prompt neutrons alone it would be 'prompt critical', and difficult to control. The dependence of the chain reaction on delayed neutrons allows you to adjust the reaction-rate gradually instead of abruptly.

Taking absorber out of a stable chain reaction is called 'adding reactivity'; the neutron density increases, and the rate of the chain reaction increases. But the build-up is gradual, because some of the neutrons do not emerge immediately after fission. The smaller the added reactivity, the longer is the time taken for the neutron density to increase by a given proportion. This time is called the 'reactor period', and is a very important measure of how well the reactor can be controlled. When a reactor has a short period it is liable to be skittish. Of course inserting absorber – 'adding negative reactivity' – produces a reverse effect. When the desired rate is established you reposition the absorbers to stabilize the reaction at that rate.

To economize on neutrons you can surround the reactor core with a reflector to bounce errant neutrons back into the reaction region. The best reflecting materials are the moderator materials; in effect you can extend the volume of moderator beyond the region of fuel elements. Since the presence or absence of reflector affects the neutron density in the core, you can add reactivity by adding reflector, or vice versa; some reactor designs utilize this effect for control purposes.

Before pulling the control rods far enough out to let your reactor go critical you must take precautions against the radiation pouring out from the core. Neither alpha nor beta particles will get beyond the fuel cladding (unless it leaks); but gamma rays and neutrons can travel through metres of concrete and still be dangerous to living matter. Therefore, you surround your reactor with enough concrete or other protective 'shielding' to cut down the radiation outside to as low a level as you think advisable.

Xenon Poisoning

Normal start-up and shutdown of a reactor are both lengthy processes, and may take many hours. If it is necessary to stop the chain reaction quickly, for instance in the event of a malfunction, the emergency shutdown is called a 'scram'. If an operating reactor is left to itself its reaction-rate will gradually dwindle, not necessarily because fissile nuclei are being used up – in some reactors the numbers of fissile nuclei may even be increasing – but also because of the build-up of fission products which absorb neutrons.

The most voracious of all is xenon-135. The consequent phenomenon, called xenon poisoning, is an intriguing demonstration of the slightly surrealistic circumstances in which reactors operate.

When you start up your reactor for the first time, the fuel contains no xenon-135. For several hours after start-up, fission processes generate tellurium-135 and iodine-135, which in turn generate xenon-135, which starts gobbling neutrons. Each xenon-135 nucleus which succeeds in capturing a neutron is thereby changed into xenon-136 – much less voracious. The xenon-135 nuclei which fail to capture neutrons nonetheless undergo beta decay into caesium-135 – also much less voracious. Accordingly, after matters have had a chance to settle down, as much xenon-135 is being lost as is being generated. There is a certain average concentration of xenon-135 in the reactor core, which remains the same as long as the chain reaction proceeds at the same rate. You budget for so many neutrons lost to xenon-135, and operate accordingly. But when you change the reaction rate you upset the balance, and the consequences may be embarrassing.

Iodine-135 turns into xenon-135 with a half-life of 6.7 hours. Xenon-135 turns into caesium-135 with a half-life of 9.2 hours – that is, slightly more slowly. Suppose you shut down your reactor. The neutron flux falls to near zero; xenon-135 stops capturing neutrons. From the moment of shutdown more xenon-135 is being generated than is being lost: while your reactor is shut down, the amount of neutron absorber in its core is steadily, surreptitiously increasing. If, several hours later, you try to start up your reactor again, you may, even with the control rods completely out, be unable to add enough reactivity to reach criticality. To be always able to start up your reactor at any time after shutdown, you will find it necessary to include more fuel, or otherwise to arrange for an excess of available reactivity over and above what normal operation needs. Apart from the obvious cost of the extra fuel, this means that even in normal operation you must leave some control rods partly inserted. It is not easy to do so without distorting the uniform neutron density in the core, and producing a less than ideal pattern of chain reaction. Reactor designers have to decide what sort of compromises they can best achieve, to satisfy the conflicting requirements made necessary by phenomena like xenon poisoning.

Refuelling

While you operate your reactor, changes take place in the fuel. The number of uranium-235 nuclei dwindles gradually as they undergo fission. Some of the uranium-238 nuclei capture neutrons and change to plutonium-239. Some of these plutonium-239 nuclei undergo fission. Others capture additional neutrons and become plutonium-240, plutonium-241 and other isotopes of elements heavier than uranium – 'transuranic actinides'. Fission products are formed; most fission products are radioactive, and undergo radioactive changes into more stable nuclei, some very rapidly, others very slowly. Fission products also capture neutrons. The composition of the reactor fuel grows increasingly complex as the chain reaction proceeds; it becomes more and more difficult to keep track of all the competing processes taking place. Some of the fission products are gaseous, like krypton and xenon; these gaseous fission products build up inside the fuel, exerting pressure and trying to leak out.

The intense neutron flux plays havoc with the crystal structure of the fuel, the cladding, and possibly the moderator, knocking the nuclei out of place and setting up stresses and strains in the material. Sooner or later, it is necessary to take out used fuel and replace it.

There is an assortment of different procedures for 'refuelling' or 'recharging' a reactor. Some designs can be refuelled while the reactor is in operation, replacing one or more fuel elements at a time: 'on-load refuelling'. Other designs are shut down for refuelling, and perhaps one-third of the core is replaced at one time: 'off-load refuelling'. All refuelling procedures must be carried out with extreme care because of the intense radioactivity of the fission products in the reactor core and in the used or 'irradiated' fuel.

Power from a Reactor

If you set up a reactor on a sufficiently large scale, and let the chain reaction run fast enough, the energy released by the rupturing uranium-235 (and plutonium-239) nuclei makes the whole assembly hot – potentially very hot indeed. Complete fission of all the nuclei in a kilogram of uranium-235 would release a total energy of about one million kilowatt-days – that is, as much heat as would be given off by one million one-bar electric fires operating for one 24-hour day. That is a lot of heat. Accordingly, the fuel in a reactor must be arranged so that the heat is given off gradually enough to keep temperatures manageable. The amount of heat given off per unit volume in a reactor core is called the 'power density'. It may be anything up to several hundred kilowatts of heat per litre; if such an outpouring of energy is not to melt – and indeed boil – the whole aggregation of material, it must be efficiently removed.

You remove the heat from the reactor by pumping a heat-absorbing fluid through the core, past the hot fuel elements. The fluid can be a gas, such as air, carbon dioxide, or helium; or a liquid, such as water or molten metal. The choice of cooling fluid – 'coolant' – depends on how fast heat must be removed; on how expensive the fluid is; on how easy it is to pump – and so on. The cooling system can be open-ended, passing ordinary air or water directly through the core and back to the atmosphere or river; such an arrangement has the virtue of simplicity, but may also have serious drawbacks, especially if fuel cladding leaks. Alternatively the cooling system can be one or more closed circuits, in which the same coolant passes through the

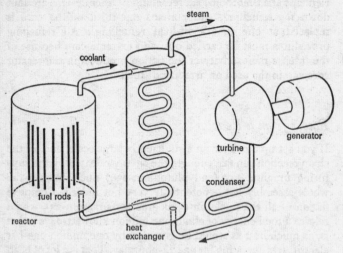

Figure 1   Nuclear power station

core again and again, carrying heat out of the core, discharging it outside the reactor, and then passing the rest of the way around the circuit and back through the core again. If the cooling system is made up of closed circuits, expensive or exotic coolants can be used, since they are confined in the system instead of being lost. A closed circuit can also be pressurized, which will in most instances dramatically improve efficiency of the coolant; a pressurized gas is denser and can carry more heat per unit volume.

The cooling system, of whatever design, removes the heat from the reactor core; what becomes of the heat thereafter depends on the reason for operating the reactor. The first large reactors were operated exclusively to generate neutrons and turn uranium-238 into plutonium-239 for nuclear weapons. The heat released in their cores was just a nuisance to be got rid of, into the near-by air or water. But with appropriate arrangements such heat, like the heat from burning coal or oil, can be used. In particular it can generate steam to run turbines or other electrical generators. Such an arrangement – a nuclear reactor providing heat to run an electrical generating plant – is called a nuclear power station, or (in the USA) a nuclear power plant.

In the following chapter we shall look much more closely at the structure and operation of the several main types of reactors. Each type releases energy by the fission of nuclei; but each uses a different arrangement of these nuclei: different designs of fuel, of moderator, of cooling system, of controls, et cetera. The differences have many important implications, as we shall see.

2. Reactor Types

With different fuels, moderators, control systems, cooling arrangements, spatial configurations and so on, possible designs of nuclear reactor number in the hundreds. Early reactor designers had a field day, letting their imaginations run riot; some of their suggestions made colleagues' hair stand on end. Others seemed more feasible: amenable to engineering, using manageable materials, controllable, safe, and – ultimately – even economic to build and operate.

As we shall see, the main lines of development of such commercial reactors sprang from the three partners in the Second World War 'atom-bomb' programme, the 'Manhattan Project'. Britain in due course developed gas-cooled, graphite-moderated reactors; the USA developed reactors cooled and moderated by ordinary 'light' water; and Canada developed reactors moderated by heavy water, variously cooled. Both Britain and the USA also began development of reactors using fast neutrons, with liquid metal coolant and no moderator. Before we describe these and other reactors in detail, it may be useful to identify some general aspects of reactor design.

To generate a given output of energy a reactor may have a very large volume of core, with a comparatively low heat output per unit volume or power density; alternatively it may have a much more compact core with a higher power density. Natural uranium reactor fuel has a low concentration of fissile nuclei; a reactor using such fuel must have a larger volume of core than one using enriched uranium – or plutonium – fuel. A large reactor costs more to build than a smaller one of the same output. On the other hand, natural uranium fuel is much cheaper than enriched uranium fuel. What you lose by building the larger, more expensive reactor you may subsequently save on fuel costs.

The energy output from a reactor can be measured directly as heat. If this heat is used in a 'power reactor' to generate electricity, only a fraction of the total heat energy ultimately reappears as electrical energy; the rest is discharged to the surroundings as low-temperature heat. In general, the higher the temperature the reactor can achieve, the larger the fraction of energy that can be converted to electricity. As a rule only some 25 to 32 per cent of the total heat output is converted to electricity in systems now operating. A system which converts 30 per cent of the heat to electricity is said to be 30 per cent efficient – mainly because the remaining 70 per cent of the heat is not used. (This is not to say that it *cannot* be used, merely that it *is* not.) Reactor energy outputs are accordingly described either as heat – for instance, 'megawatts thermal', MWt – or as electricity – for instance, 'megawatts electric', MWe. (A megawatt is one million watts.) A satisfactory rule of thumb is to assume that, for a given power reactor, the output in MWe is between one-quarter and one-third of the output in MWt. Unless use is made of the low-temperature heat, the fraction MWe/MWt is a measure of the system's efficiency.

If a reactor core operates at a higher temperature, it produces steam of higher quality, and generates electricity more efficiently. On the other hand, core materials which withstand these higher temperatures are likely to be more expensive. Similarly, reactor fuel which can be left in the core for a longer period at a higher temperature reduces the amount of fuel needed for refuelling; but such fuel also costs more. A reactor which can be refuelled 'on load' – without having to shut down – is less inconvenient for an electricity system; but such refuelling arrangements are in general more expensive to build than those for 'off-load' refuelling.

The cooling system of a reactor may operate at a pressure anywhere from atmospheric up to – at present – about 150 atmospheres. The higher the pressure, the heavier and stronger must be the pressure system. This has implications not only for costs but also for safety, since a rupture of the pressure system might have serious consequences, as we shall see. Some designs enclose the reactor core in a pressure vessel of heavy welded steel; others use prestressed concrete. Still others distribute the core materials in an array of much smaller pressure tubes.

Interruption of cooling may be easier to control in a reactor of low power density than in one of high power density, in which sharp temperature rises may occur with extreme rapidity. Flaws or malfunctions in the pressure system may be easier to overcome in a reactor with low coolant pressure than in one with high coolant pressure; a large welded steel pressure vessel of complex geometry seems inherently more vulnerable to major disruption than a pressure vessel of prestressed concrete, or a system made up of many smaller pressure tubes.

One large reactor may cost less than two small ones producing the same total output – but, as we shall discuss (pp. 231–2), not necessarily, if the large reactor must add many extra items of equipment for reasons of safety, stand-by and maintenance. All reactor designs have shared one characteristic: a rapid increase in the size of successive reactors of the same basic design. In reactors, perhaps more than in most other engineering technologies, a change of scale is frequently not merely quantitative but qualitative, introducing a whole new set of unknowns into the engineering. In Table 1 (pp. 80–83) we have listed typical design parameters of different reactor types. In the succeeding sections we shall describe them in more detail.

Experimental and Research Reactors

The first nuclear reactor was constructed amid tight wartime secrecy, in a disused squash court under Stagg Field football stadium at the University of Chicago. Construction of the reactor began in November 1942 and took less than a month. Bricks were machined out of graphite. In some of the bricks were imbedded balls of uranium metal or compressed uranium oxide powder; uranium oxide had to be used because at that time only 5600 kg of pure uranium metal were available. The graphite bricks were laid layer after layer onto a growing pile of roughly spherical configuration inside a wooden supporting structure. At intervals inside the pile were neutron-absorbing cadmium strips to ensure that stray neutrons did not initiate a premature chain reaction. Instruments to measure neutron-density were also included, and checked regularly to see how the pile was progressing towards critical dimensions.

By the time the 57th layer of bricks had been added it was clear that only the inserted neutron-absorbers were keeping the pile from criticality. By this time it was more than 6 metres high, with length and breadth to match, and contained about 36 tonnes of uranium and over 340 tonnes of graphite.

On 2 December 1942 the scientists and technicians gathered on the balcony of the squash court, watching the instrument readings, while Enrico Fermi called out instructions and a young physicist named George Weil slowly pulled out the final control rod. Shortly after 2.30 p.m. the instruments recorded a steadily-rising increase in neutron density in the pile. The pile had 'gone critical': the first self-sustaining nuclear chain reaction was taking place.

The heat generated in the pile was initially kept down to about 0.5 watts. But on 12 December the reaction-rate was allowed to increase until the heat generated – the 'power level' – reached 200 watts. Further reactivity was available, but by this power level the radiation from the pile was potentially harmful to personnel. Accordingly, in the spring of 1943, Chicago Pile No. 1 – CP-1, as it came to be called – was unpiled. Shortly thereafter, rebuilt with added uranium and graphite inside adequate radiation shielding, at a site outside Chicago, and rechristened Chicago Pile No. 2, it could be operated at an average power of 2 kilowatts (2 kWt) and intermittently up to 100 kW. For some years, until the name became totally inappropriate, any nuclear reactor was called an 'atomic pile', after the first reactor.

CP-1 was the first true nuclear reactor. But even before it was built, thirty piles of less than the necessary size and shape were built and tested. Such assemblies, which cannot generate their own neutron supply without a supplementary source of neutrons, are called 'sub-critical assemblies'. Since the early 1940s countless hundreds of sub-critical assemblies have been built and dismantled in many countries; and many true reactors have been built for experimental or research purposes. The variety and range of designs of experimental and research reactors is extensive, depending on the purpose for which a particular reactor is constructed. Experimental and research reactors have a number of uses. To bombard a sample of material with neutrons, the sample can be inserted through a suitable channel into a reactor core. The intention may be simply to study the effect of neutron bombardment on the material – perhaps a material to be used in building a reactor. Or the intention may be to convert some of the sample's stable nuclei by absorption of neutrons into radioisotopes, for medical, industrial, agricultural or research purposes. Some reactors have a 'thermal column': a graphite panel through the reactor shielding, which allows a stream of thermal neutrons to emerge for research work outside the reactor. (Although thermal neutrons are often called 'slow neutrons', their speed is nonetheless about 2200 metres per second – considerably faster than a high-velocity bullet.)

Some research reactors are designed to further the study of 'reactor physics' itself: neutron densities, temperatures, the production of plutonium-239 from uranium-238, the build-up of fission products, the effect of these fission products on reactivity, the performance of new designs of fuel assemblies, the effects of 'unscheduled events' inside the reactor, and so on.

Obviously, research reactors are also important for the training of qualified scientists and technicians in the often extremely subtle and intricate – and potentially dangerous – characteristics of reactor design and operation. Many countries now boast major centres for reactor research and development. Nor are such reactors found only in heavily industrialized countries. One of the longest-serving research reactors in the world, in operation since 1959, is the 5 kW Trico reactor in Zaïre.

A popular design of research reactor is the 'pool type': it has a core of highly enriched uranium at the bottom of a deep tank of water. The water acts as moderator, reflector, coolant and shielding. It also allows a direct view of the core while the reactor is critical; in no other reactor design is this possible. Because some of the radioactive emissions from the reactor travel faster than the speed of light in water, the water in a pool-type reactor glows with an eerie blue light called Cerenkov radiation.

The great majority of reactors, of whatever size or for whatever purpose, have been experimental, in that every new reactor modification and development has had to be designed and engineered on the basis of previous experience, which in this field is often inadequate or irrelevant or both. The US Atomic Energy Commission (see p. 124) went so far as to list *all* the reactors it licensed, for whatever purpose, as 'experimental' until 1971.

Plutonium Production Reactors

All uranium reactors produce plutonium, by neutron bombardment of uranium-238. The first large-scale reactors were built expressly for this purpose: to produce plutonium for nuclear weapons. A pilot model was built in 1943 at Oak Ridge, Tennessee. It could not be constructed on the simple building-block principle that sufficed for CP-1; it would have been not a little inconvenient to dismantle the entire reactor to recover the plutonium. Furthermore, the rate of transmutation of uranium into plutonium depends on the neutron density, which in turn depends on the rate of the chain reaction. If the reaction is fast enough to create plutonium at a useful rate, the heat generated becomes a major problem. Complete fission of all the nuclei in one kilogram of uranium-235 releases about one million kilowatt-days of energy; each uranium-235 fission is likely at most to initiate one further fission with one neutron and create one uranium-239 (and hence plutonium-239) nucleus with another. That is, to create one kilogram of plutonium-239 requires the fission of about one kilogram of uranium-235 – and the dissipation of all that heat.

Accordingly, the Oak Ridge reactor was built in the form of a cube of graphite perforated from one side to the other with parallel horizontal channels. Into these channels were slid cylindrical slugs of natural uranium clad in aluminium. When a fuel slug had been sufficiently irradiated it was pushed through the reactor, falling out of the graphite core and into a tank of water, for subsequent processing (see pp. 100–103). The fuel slugs fitted loosely in the channels, leaving room for a flow of cooling air to remove the heat from the reaction (eventually 3.8 MWt).

Even while the Oak Ridge pilot model was still under construction, work began on the first full-scale reactor, which was built on the bank of the Columbia river near the town of Richland, in Washington state. Construction of the first full-scale reactor, as tall as a five-storey building, only took from June 1943 until September 1944. By early 1945 three full-scale reactors were in operation. The entire industrial installation, named the Hanford reservation, was in due course to occupy nearly 1600 square kilometres, and include nine production reactors, plus a vast array of ancillary plant. The Hanford production reactors were similar in design to the Oak Ridge reactor; but their heat output was so intense that cooling by gas – helium was the original choice – was found too difficult. Cooling was accomplished by pumping water from the Columbia river directly through a reactor core and back into the river.

After the end of the Second World War plutonium production reactors were built in Britain, France and the USSR. The British production reactors were built on the Cumbria coast, on the site of a disused ordnance factory which was renamed Windscale. Like the Hanford reactors those at Windscale used natural uranium clad in aluminium, lying in horizontal channels in a graphite core. The absence of a suitable water supply meant that the Windscale reactors were cooled with air, blown by powerful fans through the cooling channels in the graphite, and discharged through a stack 126 metres tall back to the atmosphere. This once-through air cooling was a far from desirable arrangement, whose drawback was subsequently demonstrated in the dramatic accident in 1957 which destroyed the Windscale No. 1 reactor (see pp. 162–6).

If the purpose of a reactor is to produce fissile plutonium-239, the rate of plutonium production can be 'optimized' by choice of core geometry, at the cost of other performance characteristics. The reactor fuel must be changed at relatively short intervals – less than two years on average. By this time the fissile plutonium-239 in the fuel is playing a significant part in the chain reaction, undergoing fission and thus being removed as well as formed. Furthermore some of the plutonium-239 absorbs one or more additional neutrons without undergoing fission, becoming plutonium-240, plutonium-241, and plutonium-242. Plutonium-240, which accumulates comparatively rapidly, is susceptible to spontaneous fission, but is unlikely to fission when struck by a neutron, and cannot therefore participate in a chain reaction. It is moreover virtually impossible to separate from plutonium-239. Too high a fraction of the 240 isotope makes plutonium somewhat unpredictable as a weapons material: hence the need to remove irradiated fuel before too much plutonium-240 has been created. However, 'reprocessing' of fuel to extract plutonium (see pp. 100–103) is an expensive and complex operation. Reprocessing more often than is strictly necessary to maintain a reactor's reactivity can only be justified within the remarkable elasticity of military budgeting.

Gas-cooled Power Reactors

MAGNOX REACTORS

The first power reactors were of course, like the plutonium production reactors, military: power plants for submarines, and multi-purpose reactors producing both plutonium and electricity. (A 'nuclear submarine' is so called as much for its motive power as for its cargo.) The first 'power reactors' so identified were started up in the USA and the USSR in 1954. The US reactor had an output of 2.40 MWe, and the Soviet APS-1 reactor at Obninsk, now usually declared to have been the world's first power reactor, an output of 5 MWe.

However, for obvious reasons, the general public heard little about the first US and Soviet power reactors. By default, if not by common consent, the world's 'first nuclear power station' was Calder Hall in Britain, whose first reactor started up in 1956. Calder Hall's claim to precedence is entirely defensible, if only because the first Calder Hall reactor, like its three successors, was a full order of magnitude larger than the Obninsk reactor, with an output of 50 MWe. On 17 October 1956 Her Majesty Queen Elizabeth II switched power from Calder Hall into Britain's National Grid; in a blaze of international publicity the age of 'nuclear power' – that is, electricity, not military might – was born.

The four Calder Hall reactors, on a site adjoining Windscale, were 'power' reactors only secondarily. Despite the fanfare and the Royal première the Calder Hall reactors, and the four similar reactors built at Chapelcross across the Scottish border, were built and optimized in order to produce weapons plutonium to augment the output from the Windscale reactors. Nonetheless the Calder Hall and Chapelcross nuclear stations became the cornerstone of Britain's nuclear power programme. Their design characteristics were developed and extended through the first generation of British commercial nuclear stations, eventually comprising a total of twenty-eight reactors, including one in Italy and one in Japan. The nuclear patriarch of this family, the first Calder Hall reactor, is still going strong more than twenty years after its first start-up. Many of the factors which affected its design and construction still preoccupy nuclear engineers.

Like the Windscale reactors this Calder Hall reactor uses natural uranium fuel and graphite moderator. But their spatial arrangement is very different, as are many other details. The fundamental difference is that the Calder Hall reactor has a closed-circuit cooling system, making it possible to recover heat from the reactor at a temperature and pressure high enough to be useful. The pressurized closed-circuit system also ensures more efficient cooling, which in turn allows the chain reaction to operate and produce plutonium faster.

The heart of the Calder Hall design (see Figure 2, and Table

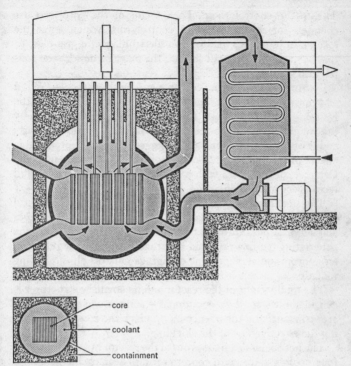

Figure 2   Magnox reactor

1, p. 80) is a huge welded steel pressure vessel, enclosing the graphite reactor core which is pierced from top to bottom by fuel channels. The Calder Hall fuel is clad, not in aluminium, but in a special magnesium alloy called 'Magnox', which is much less inclined to absorb neutrons, and is stronger and less susceptible to corrosion in the high temperature and neutron-flux inside the reactor core. The entire family of reactors using such fuel has always been referred to as Magnox reactors.

The core contains an array of instruments which transmit readings of temperatures, neutron densities and other relevant data to the control room. Each sector of the core also has channels for several types of control rods which enter the reactor from above, held out on electromagnetic grapples, so that any reactor fault will shut off the magnets and let the rods fall into the core to halt the fission reaction.

The pressure vessel, its contents and its attachments expand and contract with temperature changes. The combination of the resulting thermal stresses, the gravitational stresses set up by the weight of the components, the vibration of moving parts and fast-flowing coolant, and the somewhat unpredictable effects of prolonged intense neutron irradiation presented the Calder Hall designers with a challenge whose equivalent still faces every nuclear engineer.

The steel pressure vessel is itself enclosed inside a biological shield of concrete more than two metres thick. Assorted pipes and services pass through the biological shield; but, since the penetrating gamma rays and neutrons travel in straight lines, an appropriate arrangement of zigzags cuts off all outcoming radiation.

The total weight of the reactor and its ancillary structures is considerable – some 22 000 tonnes – and the site requirements are stringent; any subsidence might crack the concrete, reducing the effectiveness of the shielding.

The hot coolant gas passes out of the reactor building through four cooling ducts into four towering 'heat exchangers' – a fancy word for boilers. Inside each is a labyrinth of tubing containing water; the hot carbon dioxide passes round the tubing, giving up its heat to the water which turns to steam and is used to drive turbogenerators. When the gas has given up its useful heat it emerges from the lower end of the heat exchanger, and passes into a gas circulator. This blows it back into the bottom of the reactor pressure vessel and up again through the fuel channels.

Since the four loops of the cooling circuit are pressurized, special provisions have to be made for changing fuel elements, and for other maintenance inside the reactor core. Access to the channels, for refuelling and servicing, is from above, through holes in the horizontal roof of concrete shielding which is called the 'pile cap'. On the pile cap, the working area above the reactor, are mobile 'charging' or 'refuelling machines', massive and complex assemblies.

The Calder Hall Magnox design – primarily for plutonium production – is shut down and depressurized for refuelling. However, in the newer Magnox designs for commercial nuclear stations, it is not necessary to interrupt the operation of the reactor for refuelling; it can be carried out continually, a few channels per week, while the reactor is supplying power, 'on load'.

To change the fuel in the reactor, the 'discharge machine' is positioned over an access port, clamped onto the surface of the pile cap and pressurized. The shielding plug is removed, grapples extended down through a standpipe into the core, and the irradiated fuel elements lifted out of a channel and stored inside the thick walls of the discharge machine – all by remote control, because of the radiation hazards. The shielding plug is replaced, the discharge machine depressurized and moved, and the charge machine, loaded with fresh fuel, moved into position. The whole cycle is repeated, again by remote control, to lower new elements into place: clamping, pressurization, unplugging, replugging, depressurizing and unclamping.

In either case, the irradiated fuel, intensely radioactive with fission products, is moved, inside the discharge machine, to be dropped into a 'cooling pond': a deep tank of water, which serves to shield and cool the fuel while the more short-lived fission products within it decay to a less dangerous level of activity. After a suitable interlude – usually 150 days – the irradiated fuel is transported to Windscale for 'reprocessing' (see pp. 100–103).

In all, eight Magnox stations, each with two identical reactors, were built for the Central Electricity Generating Board (CEGB), and one for the South of Scotland Electricity Board. Design details varied considerably from station to station, although all incorporated on-load refuelling by means of a single charge–discharge machine. The Berkeley station's reactors use cylindrical steel pressure vessels, whereas the Bradwell station's reactors, built at the same time, use spherical ones. The Hunterston station is refuelled not from above but from below, where the temperature is lower. The different stations have different arrangements of heat exchangers and generating sets, different reactor buildings and so on.

Perhaps the most important variation in the Magnox stations is the power rating, which was increased progressively. To accommodate this increase in power and size, the last two CEGB stations introduced a major modification in design. Welding of a steel pressure vessel of more than a certain size to the stringently high standards necessary for a reactor becomes prohibitively difficult. Accordingly, the CEGB's Oldbury station embodied an entirely new approach. The pressure vessel was fabricated not from welded steel but from pre-stressed concrete, a much more manageable material for large and complex structures. In the Oldbury design not only the reactor core but also the heat exchangers and gas circulators are enclosed within the concrete pressure vessel. The pre-stressed concrete serves both as pressure vessel and as biological shield; the gas ducts are completely eliminated, removing one of the major escape routes for radioactivity in the event of an accident. The pre-stressed concrete design made possible a more than twofold increase in reactor size for the final CEGB Magnox station, at Wylfa in Wales.

The power density of the Magnox stations, which averages about 0.9 kilowatts per litre, is, by nuclear standards, low. Because of its low density and consequent low thermal capacity, gas is a less efficient coolant than liquid; accordingly, the rate of heat-generation in a gas-cooled core must be kept low. (This in turn imposes an overall limit on maximum feasible heat output, since high output entails a very large volume of core, with accompanying engineering complications.) Another characteristic of interest is the 'specific power': power generated per unit mass of fuel. The specific power of Calder Hall fuel is about 2.40 kilowatts per kilogram of uranium; that of Wylfa fuel is about 3.16 kilowatts per kilogram of uranium. Specific power is also sometimes called 'fuel rating'. The 'burn-up' of fuel is the cumulative heat output per unit mass; it is commonly measured in megawatt-days per tonne of uranium. The burn-up is of course a measure of how many fissions have occurred inside a given amount of fuel.

One of the main objectives of fuel designers is to achieve higher burn-up – that is, to be able to leave fuel in a reactor longer, before it becomes too distorted and too burdened with fission products to function properly. The limitations on burn-up of Magnox fuel are numerous. Natural uranium metal has a complicated crystal structure, and undergoes a variety of unwelcome changes at high temperatures and intense neutron fluxes. A burn-up of between 3000 and 4000 megawatt-days per tonne of uranium is about the best that can be comfortably attained by Magnox fuel. This limitation was one of several factors which eventually terminated the Magnox programme and provoked a search for another approach.

The only other major nuclear power programme to opt for gas-cooled reactors was the French. The small French power reactors at Marcoule and Avoine started up in 1958. The second unit at Avoine – Chinon-2, a 200-MWe reactor – went critical in 1964, and France has since built, in all, seven gas-cooled power reactors with graphite moderator and a 70-MWe gas-cooled reactor with heavy water moderator. But French interest has subsequently swung markedly away from gas-cooled to light water designs, in partnership with American reactor-builders (see p. 189).

ADVANCED GAS-COOLED REACTORS (AGRS)

Even while the first Magnox stations were barely under construction, work commenced on a second-generation design of gas-cooled power reactor: the advanced gas-cooled reactor (AGR). The aim was to achieve higher gas temperatures to improve the efficiency of electricity generation; higher fuel ratings to make the reactor more compact; and higher burn-up to reduce the frequency of refuelling. The temperatures achievable with Magnox fuel are limited by the characteristics of Magnox alloy and of uranium metal. Uranium metal undergoes a crystalline change of phase at 665°C, accompanied by marked expansion; its behaviour even below this temperature is complex, since it expands at different rates in different directions with increasing temperature. The melting point of Magnox is about 645°C; as well as melting at this temperature Magnox may also catch fire.

Accordingly, higher-temperature fuel must use some other form of uranium. The form most commonly chosen is uranium dioxide, UO_2, often just called uranium oxide. Whereas uranium metal melts at 1130°C, uranium oxide melts only at 2800°C. However, uranium oxide has a low thermal conductivity, much lower than that of uranium metal. When uranium metal is undergoing a fission reaction, its high thermal conductivity means that the temperature is more or less uniform through the whole thickness of a fuel rod, even if this is several centimetres. The same is not true of uranium oxide. If solid uranium oxide is undergoing fission, the heat generated in the interior does not readily make its way to the surface; the interior is much hotter than the surface. Uranium oxide fuel elements must have a smaller diameter than metallic uranium elements, even though the melting point of uranium dioxide is so much higher.

The basic building block of uranium oxide fuel is usually a pellet made by compressing, baking or otherwise persuading uranium oxide powder to assume the form of a hard small cylinder, about the size of a liquorice allsort. A column of such pellets, anything up to several metres in length depending on the fuel design, is stacked inside a thin-walled metal tube, to make a 'fuel pin'. The tube must be of a material which can withstand high temperature. Some oxide fuels use an alloy of zirconium, which has advantages but is expensive; the material most commonly chosen is stainless steel, as is the case with fuel for the advanced gas-cooled reactor. Stainless steel involves a further problem; it is strong and well-behaved structurally, but it has an unhealthy appetite for neutrons. Accordingly, the percentage of uranium-235 in the uranium oxide must be increased above its natural level: that is, the uranium oxide must be enriched (see p. 91–5). In AGR fuel the uranium is usually enriched to about 2 per cent.

The first AGR incorporating this type of fuel was a small prototype built at Windscale. The Windscale AGR started up in 1962; as this is written it is still the only AGR ever to have gone critical. Five full-scale stations, each with a pair of twin AGRs, have been under construction since the late 1960s; the station designs differ, but all have given problems, some of them acute (see pp. 183–4).

The basis of the overall AGR design is the prestressed concrete pressure vessel first developed for the last two Magnox stations (see Figure 3, and Table 1, p. 80). Like a Magnox reactor, an AGR has a core of machined graphite, under a dome like a huge steel bell-jar, with a large number of openings at the top, through which pass the standpipes for access to the fuel channels. Outside the dome – but still inside the pressure vessel – are the heat exchangers or boilers, and below them the gas circulators.

The amount of fuel in an AGR is considerably less than in a Magnox reactor of comparable output, while the fuel rating is considerably higher. The coolant gas emerges from the fuel channels at a temperature of around 650°C, more than 300° higher than normal Magnox operating temperature.

An AGR is refuelled by a single refuelling machine, which pulls an entire fuel string of eight elements out of the reactor at once. Accordingly, the refuelling machine is itself the height of a four-storey building, and the reactor building must be built like an aircraft hangar to accommodate it. A single machine serves both reactors at a given station, moving between them, the fuel store and the spent fuel pond on a gantry or rails.

HIGH TEMPERATURE GAS-COOLED REACTORS (HTGRS)

Anyone desiring a heat source will have two objectives in mind: the total amount of heat output per unit time (that is, the total power) and the temperature at which the heat is made available. There is an unimaginable amount of heat in the ocean; but its

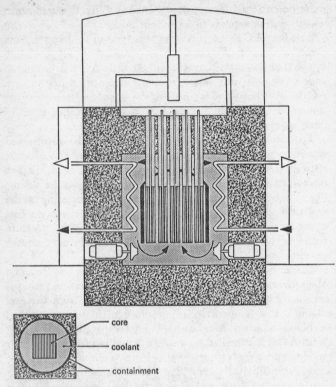

core

coolant

containment

Figure 3   Advanced gas-cooled reactor (AGR)

low temperature makes it of little overt use. While reactor designers were scaling up reactor sizes, to increase their power output, they were also pressing on towards much higher temperatures. Even an AGR operating flat-out is only a so-so source of heat, as far as temperature is concerned. It can be used to raise passable steam to run a turbogenerator and produce electricity, but only at a moderate efficiency. More elegant industrial applications are ruled out by the low temperature of the heat.

The limitation on the temperature at which heat is generated is nothing – or almost nothing – to do with the chain reaction system itself. Under the right circumstances a chain reaction can run at temperatures anywhere up to those in the heart of a nuclear explosion – millions of times higher than those in fossil-fuel boilers. However, long before such temperatures are reached it becomes peculiarly difficult to keep the whole assemblage in any semblance of order. We have already noted the awkwardness of uranium metal and Magnox cladding at temperatures over 600°C; other reactor materials present similar problems, albeit at variously higher temperatures. Clearly if really high temperatures are to be permissible, without having the entire reactor core bulge and warp itself hopelessly out of shape, or undergo unpleasant chemical reactions, a new approach is necessary. The new approach which has received the most attention is one which dispenses entirely with metals in the reactor core, in favour of sophisticated combinations of refractory ceramic materials able to withstand without protest temperatures well into the thousands of degrees Centigrade.

Work began in 1957 on the first two such reactors. Under the aegis of the forerunner of the Organization for Economic Cooperation and Development (OECD) an international project was established at Winfrith, Dorset, England to build the Dragon High Temperature Gas-cooled Reactor; in the USA the General Atomic Company began the programme which led to the construction of the Peach Bottom-1 reactor near Philadelphia, the world's first HTGR power station. The Dragon reactor started up in 1964, Peach Bottom-1 in 1965 and another small HTGR, the AVR, near Jülich, West Germany, in 1966. Larger stations have since been undertaken in the USA and West Germany; thus far only one, at Fort St Vrain, Colorado, has gone critical, in early 1974.

The various designs of HTGR are, like Magnox reactors and AGRs, cooled by a gas and moderated by graphite (see Figure 4, and Table 1, p. 81). But at that point the resemblance abruptly ceases. Instead of having the fissile material and graphite essentially segregated, HTGR core designs involve an intimate marriage of fuel and moderator. The fissile material of

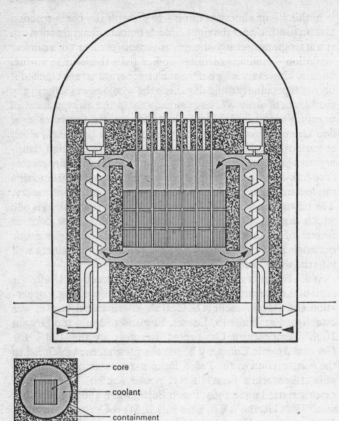

Figure 4   High-temperature gas-cooled reactor (HTGR)

HTGR fuel is – at least initially – uranium oxide or carbide, highly enriched with up to 93 per cent uranium-235, formed into tiny spheres. The spheres are coated with one or two layers of refractory carbon and one of silicon carbide and embedded in graphite.

Some designs of HTGR fuel also include similar coated particles containing not uranium but another element, thorium. Thorium has nuclear properties rather like those of uranium-238. Natural thorium is almost entirely thorium-232. A nucleus of thorium-232 can absorb a neutron to become thorium-233, which then emits two beta particles to become uranium-233, which is fissile. The process is directly analogous to that by which uranium-238 is transformed into plutonium-239. Uranium-233, like uranium-235, undergoes induced fission when struck by a slow neutron – and in turn produces more neutrons, to sustain a chain reaction. Uranium-238 and thorium-232, although not fissile materials, are called 'fertile' materials because they can be transmuted by neutron bombardment into the fissile materials plutonium 239 and uranium 233.

The fertility of thorium is increasingly becoming of interest. When uranium–thorium HTGR fuel is first permitted to undergo a chain reaction all the fissions are of course in uranium-235 nuclei. However, the thorium is gradually transmuted into uranium-233; and it happens that uranium-233 nuclei emit more neutrons per fission on average than either uranium-235 or plutonium-239. Before long the uranium-233 in the thorium particles is making a substantial contribution to the total neutron density in the reacting fuel – and also, of course, to the total heat output of the core. All the current HTGR designs are intended to operate on the thorium cycle. The uranium-233 remaining in used fuel can be recovered by reprocessing (see p. 102) and incorporated in subsequent fuel elements.

The actual core geometry of different HTGRs varies, but the overall effect is generally to achieve a much more uniform distribution of fissile material and moderator through the core. The Dragon core, a hexagon only 1.6 metres tall, is not very large. Compared to a Wylfa core, or even to a Calder Hall core, it is miniscule. It is nevertheless capable of producing 20 MW of heat, at a specific power of 1.5 MW per kilogram. Dragon fuel burn-up is as high as 100 000 MW-days per tonne, compared to about 3000 MW-days per tonne for Magnox fuel.

Despite the preponderance of carbon in the core, the coolant used in HTGRs is not carbon dioxide but helium.

A curious feature of the HTGR is its 'negative temperature coefficient of reactivity'. This mouthful means, simply, that when the temperature rises the reactivity goes down. One reason is that thermal expansion moves the fissile nuclei farther apart. Not all reactor designs behave this way throughout their operating range. For some reactors, in some circumstances, a rise in core temperature produces an increase in reactivity. If such is the case, an increase in temperature tends to encourage itself. If the original increase in temperature is unintended the last thing the reactor operator wants is to have it provide some more reactivity of its own. Clearly, a reactor which is exhibiting a 'positive temperature coefficient of reactivity' is liable – if you will pardon the expression – to overreact. Accordingly, the comprehensively negative temperature coefficient of reactivity is an attractive feature of the HTGR.

A radically different design of HTGR is under construction at Schmehausen in West Germany. In this design, called a 'pebble-bed reactor', the core is a huge bin, filled with ceramic fuel shaped into what look like outsize black billiard balls. The balls incorporate both the fissile material and the carbon moderator. Helium coolant is blown down through the binful of balls. As the reactor operates the balls are shaken slowly down the bin, emerging at the bottom. If they have reached their irradiation limit they are discharged; otherwise they are returned to the top of the bin for another trip.

The opportunity for increased efficiency of steam-raising for electricity generation is of itself a recommendation for HTGRs of whatever design. But the possibility also exists of using an HTGR in a direct cycle, passing the hot helium directly through a gas turbine, eliminating the steam cycle entirely. Another application now receiving increasing attention is to use an HTGR to generate process heat for high-temperature industrial processes like steel-making. It is even reported that Soviet researchers are making progress toward the direct transformation of HTGR heat into electricity, on a practical scale.

Using a bank of silicon–germanium units, they are said to be able to produce a steady output of 10kW, by a scaled-up thermoelectric effect alone, with no moving parts.

Light Water Reactors

PRESSURIZED WATER REACTORS (PWRS)

Like the first British power reactors, which were built to produce weapons-plutonium, the first US power reactors also began under military auspices, albeit specifically as power plants. The US Navy realized after the Second World War that a submarine powered by nuclear fuel would not need to resurface to replenish oxygen supply, since the 'burning' of nuclear fuel – unlike that of oil – does not require oxygen. Spurred by this idea, and constrained by the space limitations in a submarine, US designers developed a reactor using a core of relatively high power density, with fuel elements immersed in a tank of ordinary water – called 'light water' to distinguish it from heavy water – under sufficient pressure to keep it from boiling. The 'first power reactor ever built', according to its builders, went critical on 30 March 1953 in a land-based mockup of a submarine hull at the National Reactor Testing Station in Idaho. The following year saw the launching of the uss *Nautilus*, the first nuclear-powered submarine, powered by a pressurized water reactor system. In 1957 the submarine reactor came ashore, as the Shippingport power station near Philadelphia, the first nuclear power station in the USA. In subsequent years the pressurized water reactor, or PWR, has become the world's most popular. As this is written there are 128 PWRs in operation or under construction in 17 countries, and many more planned.

The basic structure of a PWR (see Figure 5, and Table 1, p. 81) is a large pressure vessel of welded steel with a lid held onto the upper end by a ring of heavy bolts. The pressure vessel contains the reactor core, and other so-called 'reactor internals' like control rods; the remaining volume is completely occupied

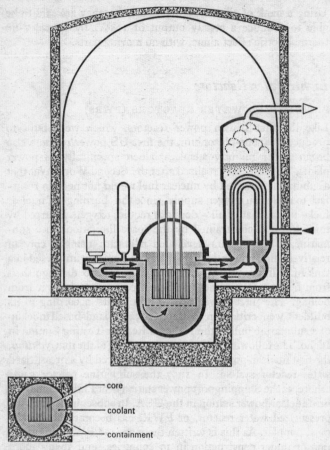

core

coolant

containment

Figure 5   Pressurized water reactor (PWR)

by ordinary 'light' water under a pressure of about 150 atmospheres. The core is made up of fuel elements, each a faggot of full-length fuel pins. A PWR fuel pin is a tube of a zirconium alloy called zircaloy, about 1 centimetre in diameter, filled with stubby cylindrical pellets of uranium dioxide. So far as neutrons are concerned the zircaloy cladding is comparatively well-behaved, much more so than stainless steel – albeit more expensive. But the water in which the whole concatenation is immersed is – as noted in Chapter 2 (p. 34) – an enthusiastic gobbler of neutrons, and to offset its distracting influence the uranium in PWR fuel pellets is enriched to about 3 per cent uranium-235. The water inside the pressure vessel serves simultaneously as moderator, reflector and coolant. At the top of the core it leaves through heavy pipes welded to the pressure vessel. PWRs can have two or more 'loops' of cooling circuit. In each loop, the pipe through which the water enters the pressure vessel is called the 'cold leg', and that through which it leaves is called the 'hot leg'. Some of the most intense discussion in recent years has focused on the consequences of a postulated break in a cold leg of a PWR (see pp. 193–8).

The hot leg of a PWR cooling loop carries the hot coolant water into a steam generator or boiler. The hot water from the reactor passes through thousands of tubes immersed in more water, under considerably lower pressure. Although the pressurized water inside the tubes cannot boil, the lower-pressure water outside them does. The resulting steam is processed and piped to a turbogenerator set. The primary coolant water returns through the cold leg to the reactor vessel, encouraged by a primary coolant pump. Each coolant loop also includes a 'pressurizer', in which an appropriate quantity of the coolant water is evaporated or condensed, to maintain coolant pressure and to compensate for the effects of thermal expansion and contraction as plant output varies. The pressurizer can also help to offset unintended increases in system pressure resulting from malfunctions. The electric immersion heaters in a pressurizer can generate 2000 KW – a bit overwhelming for a household hot-water system.

Easily the most controversial feature of the PWR are the emergency core cooling systems, provided to prevent overheating of the reactor core in the event of an accident. However, rather than describing them here, it will be more appropriate to defer their description to Chapter 7 (pp. 193–8); there can be

T–C

few technologies which have been subjected to such exhaustive – and inconclusive – scrutiny.

PWR control and instrumentation systems vary widely in design. But control rod assemblies are commonly suspended above the core, inside the pressure vessel lid, with drive mechanisms functioning through the lid from above. A PWR is refuelled off load – that is, with the reactor shut down. The reactor is allowed to cool. Then a pool-shaped chamber above the reactor – the 'reactor well' – is flooded with water, to provide shielding and cooling; the lid is unbolted and moved to one side, exposing the interior of the reactor. Since the whole procedure is time-consuming, a substantial portion of the fuel charge is changed at each refuelling – typically about one-third of the core. PWR designers usually provide for one refuelling operation annually.

Needless to say a PWR is, like any power reactor, enclosed in heavy shielding. The reactor vessel itself is surrounded by two or more metres of concrete, extending upwards to form the side walls of the reactor well. It is usual to provide a certain amount of shielding for the entire primary circuit – steam generators, primary pumps, pressurizers and piping – because the primary coolant is commonly slightly radioactive (see pp. 104–6). The reactor building itself is usually designed to serve as a secondary containment.

Some PWRs deliver close to 4000 MW of heat at a power density over 100 KW per litre. But the low coolant temperature attainable using water under manageable pressure – some 150 atmospheres, as noted – makes the PWR a comparatively inefficient source of heat for electricity generation. Nonetheless it continues to find eager customers.

BOILING WATER REACTORS (BWRS)

US interest in water cooling of reactors stemmed from the Hanford reactors and was furthered by the submarine PWRs. It was known that water allowed to boil is more effective in removing heat, but boiling was thought likely to trigger instabilities in a reactor core. The water in such a core serves also as moderator; if a steam bubble forms, the local effect on reactivity is swift and its consequences difficult to predict. But experiments in the mid 1950s demonstrated that water could indeed be allowed to boil in a reactor core. Accordingly, a new design of reactor was developed, which is by far the simplest in concept of all power reactors: the boiling water reactor, or BWR (see Figure 6, and Table 1, p. 82).

BWRs and PWRs are often mentioned in the same breath, as 'light water reactors' or LWRs. In a BWR the water serves as moderator, reflector and coolant – and in addition, when boiled, produces steam which is ducted directly to drive a turbogenerator. Once through the turbines, the coolant water is condensed and pumped again into the 'boiler' – that is, the reactor pressure vessel.

The pressure which the vessel must contain need not be much more than the pressure of the steam being produced – usually less than half that in a PWR. Accordingly, the pressure vessel need not be so thick. A BWR pressure vessel also includes the whole steam-collection and processing array, above the core. The control rods therefore enter a BWR core from below. The cooling circuits of a BWR bear little resemblance to those of a PWR. In a BWR water boils inside the fuel assemblies, and there are no external steam generators. The consequent saving in capital cost is a major factor in the hot competition between sellers of PWRs and BWRs.

Since a BWR is coupled directly to the turbine of a generating set, special provision must be made to dispose of steam if the turbogenerator cannot for any reason accept it, or if any malfunction should occur. A BWR is therefore enclosed – pressure vessel, attached piping and all – inside a 'primary containment', which consists of a huge flask-shaped concrete housing called, confusingly, a 'dry well'. Cavernous pipes lead from the bottom of the dry well down into a ring-shaped tunnel, amply large enough to walk through, half-filled with water. This tunnel is called a 'pressure suppression pool'. If for any reason steam or water escapes from the reactor vessel or the pipework, it is

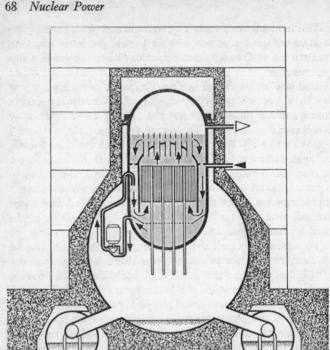

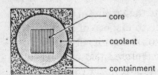

core

coolant

containment

Figure 6   Boiling water reactor (BWR)

confined in the dry well and channelled down through the pipes leading into the water in the pressure suppression pool. Any steam which gets this far is thereupon condensed, and any excess pressure it would otherwise exert on the containment is – as the name suggests – 'suppressed'.

The function of the BWR containment is closely associated with that of the emergency core cooling systems, provided, like those in a PWR, to prevent overheating of the reactor core in the event of an accident. Once again, we shall defer further description of these features until Chapter 7 (p. 193–8).

Like a PWR, a BWR is refuelled off load, with the reactor shut down and cooled. Refuelling of a BWR is somewhat more of a chore; as well as flooding the reactor well, and unbolting and removing the lid, it is also necessary to lift out and set aside a motley assortment of steam-processing fittings.

Like the coolant in a PWR, the coolant in a BWR may become slightly radioactive. Since the primary coolant in a BWR supplies steam directly to a turbine, some of the radioactivity in the coolant may reach the turbine. However, in practice most of the radioactivity in BWR coolant stays in the liquid water, and does not travel with the steam to the turbine.

The BWR shares with the PWR the drawback of comparatively low coolant temperature, and resulting inefficiency of conversion of heat to electricity. A typical BWR output temperature is less than 300°C. On the other hand the BWR also shares with the PWR the problems associated with relatively high power density, as we shall discuss further in Chapter 7 (pp. 193–8). The BWR is also more susceptible to 'burn-out' or 'steam blanketing', which arises if a layer of steam forms next to the hot fuel cladding. The low heat conductivity of the steam means that the heat is no longer so effectively removed from the fuel, and the fuel temperature may rise suddenly and dangerously.

Design and operation of all types of reactor must take into account the possibility of sudden surges, called transients: temperature transients, pressure transients and so on. This is particularly true for reactors of high power density, like the light water reactors.

Heavy Water Reactors

CANDU REACTORS

The Canadian role in fission research during the Second World War was particularly concerned with heavy water. But after the war Canada decided against embarking on a nuclear weapons programme. Accordingly, with no facilities for uranium enrichment, but with a plentiful supply of indigenous uranium, Canada chose to concentrate on heavy water natural uranium reactors. For some years efforts were directed primarily to fundamental research; during this period, in 1952, a Canadian heavy water research reactor, the NRX at Chalk River, suffered the world's first major reactor accident (see pp. 159–61). By the mid 1950s interest began to focus on the development of a power reactor, indeed a family of power reactors, sharing the family name CANDU (for CANadian Deuterium Uranium). The name also serves as a trade-mark, echoing the North American assertion of capability, 'can do'.

In 1971 the CANDU design came of age with the start-up of the first and second of four 508-MWe reactors at Pickering, near Toronto. The Pickering station, with all four reactors in operation, has been called the world's largest nuclear power station; a still larger one, using still larger reactors, is now under construction at Bruce on Lake Huron.

The design used at Pickering and Bruce is called the CANDU-PHW, since it uses pressurized heavy water as coolant (see Figure 7, and Table 1, p. 82). The heart of the CANDU-PHW is a horizontal cylindrical stainless steel tank, with circular ends. Through this tank, called the 'calandria', run horizontal zircaloy tubes. Inside each of these calandria tubes is another similar tube of slightly smaller diameter; this inner tube is a pressure tube inside which lie twelve short bundles of fuel rods. The fuel rods, natural uranium oxide pellets in zircaloy tubes, form a cylindrical faggot, containing 22 kilograms of uranium oxide. The space in the pressure tube not filled by fuel bundles is taken up by heavy water, flowing

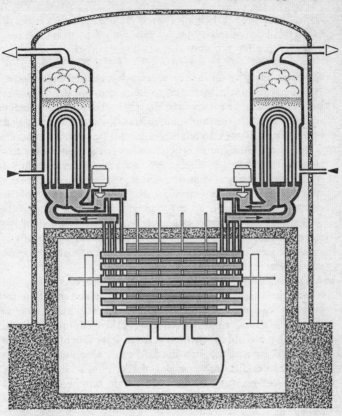

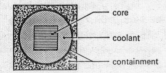

core

coolant

containment

Figure 7 CANDU reactor

through the tube. Emerging from individual pressure tubes at each end of the calandria, the hot heavy water feeds into larger-diameter 'header' pipes which carry it to steam generators.

In a graphite-moderated reactor the core can be made out of solid graphite with holes drilled through it for fuel and coolant. It is not easy to drill permanent holes through heavy water; but comparable geometry is achieved by containing the heavy water moderator in a tank – the calandria – shaped *as though* it had horizontal holes drilled through it for fuel and coolant.

The moderator circuit is kept cool and at atmospheric pressure, and the space not filled with liquid heavy water is occupied by helium cover gas. Below the reactor core is a dump tank which can accommodate the entire heavy water inventory of the moderator system.

Control rods enter the reactor from above. Only one of the eleven shutdown rods passes between any two calandria tubes; a mechanical distortion of a tube in the event of accident would jam at most two of the eleven rods.

The refuelling system of a CANDU reactor is complex and ingenious. A CANDU is designed to be refuelled continuously on load. The arrangement recalls the earliest plutonium production reactors, although the CANDU technique is much more elaborate and fully automated. At either face of the reactor is a refuelling machine in a shielded vault. One machine rams fresh fuel bundles into one end of the tube, while the other collects used ones as they emerge from the other end. The full machine then feeds the used fuel down a conveyor to eventual storage in a large water-filled cooling pond under the station.

The cooling pond at the Pickering station has storage capacity for ten years' output of used fuel from the four reactors at the station operating at full output. At present the used fuel is simply stored, rather than being reprocessed; Canada does not have commercial reprocessing facilities (see pp. 100–103). But each fuel bundle contains valuable plutonium – the heavy water natural uranium design is especially well suited to the production of plutonium. In due course it will be necessary to decide what to do with the accumulated used fuel bundles.

Variations of the basic CANDU design include the White-shell Reactor WR-1, with an organic fluid coolant, which might lend itself well to a thorium fuel cycle; and the Gentilly-1 CANDU-BLW (for Boiling Light Water), in which a light water coolant is allowed to boil in vertical fuel channels, the steam passing in a direct cycle to a turbogenerator.

STEAM GENERATING HEAVY WATER REACTORS (SGHWRs)

The British Steam Generating Heavy Water Reactor (SGHWR) combines features of the CANDU-BLW and the BWR. Only a 100-MWe prototype of the SGHWR exists, at Winfrith in Dorset; it started up in 1967. The SGHWR is rather like a forest of very narrow BWRs embedded in heavy water (see Figure 8, and Table 1, p. 83). The basis of the SGHWR is a calandria filled with heavy water moderator pierced by vertical channels; in each channel is centred a zircaloy pressure tube containing one fuel element. Light water coolant flowing through these pressure tubes is allowed to boil, as in the core of a BWR, generating steam which is fed directly to a turbogenerator.

SGHWR fuel elements contain uranium oxide enriched to about 2 per cent uranium-235. Use of enriched uranium is the major difference between the SGHWR and CANDU-BLW and makes the SGHWR much more compact. The power density in its core is just over 30 kW per litre.

The power level of the Winfrith SGHWR is varied by vary-ing the level of the moderator in the calandria; if heavy water is allowed to flow out of the tank the reactor power drops, because the absence of moderator around the upper part of the pressure tubes inhibits the fission reaction. Swift shutdown or scram is achieved by squirting boron solution into special channels in the core.

The SGHWR is a modular design, which can be enlarged simply by adding fuel channels; it is therefore unusually adapt-

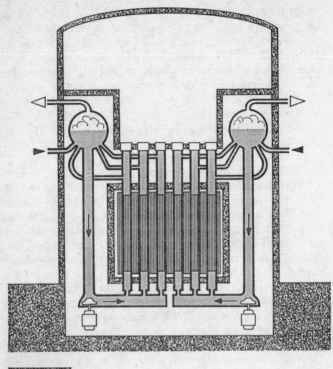

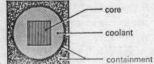

core

coolant

containment

Figure 8   Steam generating heavy water reactor (SGHWR)

able to a range of sizes. The highly efficient heavy water moderator, enriched uranium and pressure tube design also make it possible to build an SGHWR smaller than is feasible for most other designs of power reactor.

Fast Breeder Reactors (FBRS)

All the reactors so far described share a common feature. Their physical basis is fission induced by slow 'thermal' neutrons. Such reactors can be called, as a group, 'thermal' reactors. Even in a thermal reactor some of the available neutrons are absorbed by uranium-238, turning it into plutonium-239, which may then fission and make a significant contribution to the total release of energy. But the amount of plutonium created is less than the amount of uranium used up; so such reactors can also be called 'burner' reactors.

We indicated earlier that uranium-238 in this context is a 'fertile' material. In a reactor containing both fissile and fertile material, the comparison between fissile nuclei consumed and fertile nuclei converted to fissile is called the 'conversion ratio'. For instance, if for every 10 uranium-235 nuclei undergoing fission 8 uranium-238 nuclei are converted to plutonium-239, the conversion ratio is 0.8.

In a burner reactor, by definition, the conversion ratio is less than 1. A substantial conversion ratio even when less than 1 is handy. In a CANDU reactor, for instance, before a fuel bundle is discharged, an impressive number of uranium-238 nuclei have already been converted into plutonium-239 and subsequently undergone fission, making a sizeable contribution to the total heat output from the bundle. CANDU designers consider this once-through approach a particularly elegant way to utilize plutonium.

It is also possible to design a reactor with a conversion ratio greater than 1: a 'breeder' reactor, which produces more fissile material than it consumes. At the end of its sojourn in the core, fuel from such a reactor emerges containing more fissile nuclei than it contained when new. Of course it also contains the usual complement of ferociously radioactive fission products; recovering the new plutonium is not easy. Nonetheless the concept of the breeder plays a major role in present planning by the nuclear industry.

The design criteria for a breeder are very different from those which govern the reactor types thus far discussed. As mentioned earlier a thermal neutron is much more likely to rupture a uranium-235 or plutonium-239 nucleus than is a fast neutron fresh from a fission event; hence moderators are used in all burner reactor cores to slow neutrons down. This may suggest that fast neutrons are pretty ineffectual in a chain reaction. But a fission caused by a fast neutron produces on average more new fast neutrons than does a fission caused by a thermal neutron.

Breeding new fissile nuclei in a chain reaction requires, under ideal and unattainable circumstances, exactly two new neutrons from each fission: one to carry on the chain reaction by causing a further fission, and one to transmute a fertile nucleus into a fissile one. (Under such circumstances the conversion ratio is exactly 1 – replacement value.) In fact neutrons are lost to the system by leakage and by 'parasitic absorption' in coolant, reactor structure et cetera. Accordingly, to achieve a measurable rate of breeding, the reacting system must depend on fissions which produce significantly more than two neutrons per neutron lost. The most obvious combination available is fission of plutonium-239 by fast neutrons. Fission of uranium-235 with fast neutrons is less efficient, but will work. So will a mixture of uranium-235 and plutonium-239. In each case some fertile uranium-238 must be included. A reactor which breeds more fissile material than it consumes, by using a reaction dependent on fast neutrons, is called a 'fast breeder reactor', or FBR.

(An even better system is offered by using thorium-232 as a fertile material to produce its fissile cousin uranium-233; work is progressing on such a system. Its virtues have already been outlined in the section on high-temperature reactors, though without noting its potential for true breeding. However the uranium–plutonium fast breeder systems have thus far received the great majority of attention and development.)

As it happens, the first reactor ever to power electric generating equipment was a fast breeder reactor. On 20 December 1951, at the National Reactor Testing Station (NRTS) in Idaho, the Experimental Breeder Reactor-1 (EBR-1) produced enough electricity to light four 25-watt bulbs. (Four years later, the EBR-1 suffered an accident that melted the material in its core (see pp. 167–8).) The first true power reactors based on the fast breeder principle were the British Dounreay Fast Reactor in Caithness on the north Scottish coast, the Experimental Breeder Reactor-2 (EBR-2) at NRTS, Idaho, and the Detroit Edison Enrico Fermi-1 reactor near Detroit, Michigan. The Dounreay Fast Reactor has been in operation since 1959 and the EBR-2 since 1963. However, the Detroit Edison reactor, which was intended to be the prototype of a full-fledged commercial fast breeder reactor, experienced endless trouble, including an accident that might have necessitated the evacuation of Detroit (see pp. 180–3). It has now been permanently shut down and is being dismantled or 'decommissioned'.

A new generation of prototype fast breeder power reactors has now emerged, including the Soviet BN-350 reactor at Shevchenko on the Caspian Sea, the French 250-MWe Phénix reactor at Marcoule, and the British 250-MWe Prototype Fast Reactor (PFR) at Dounreay, all of which have gone critical since November 1972. The Fast Flux Test Facility at Hanford is still under construction after long delays; but the USA is now embarking on a programme which is to begin with a demonstration 380-MWe fast breeder power station at Clinch River in Tennessee.

The fundamental difficulty facing the designer of a fast breeder reactor is that it takes 400 times as many fast neutrons as thermal neutrons to cause a fission. Accordingly, a much higher neutron density must be created. Furthermore, newly emergent neutrons must avoid collisions which would slow them down before they strike other fissile nuclei. The core of a fast breeder reactor must thus be far more compact than that of any power reactor thus far described. Not only does it contain no moderator: it also contains a minimum of other structural material, and as little coolant as suffices to carry away a fiercely intense output of heat. The technological challenge is one of the most demanding ever encountered.

The overall layout of a fast breeder is a compact core of

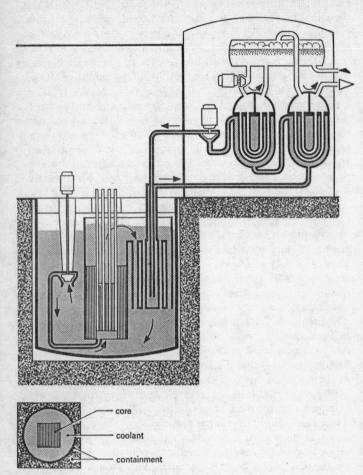

Figure 9   Fast breeder reactor (FBR)

concentrated fissile material surrounded by a 'blanket' of fertile material to catch the neutrons pouring from the core (see Figure 9, and Table 1, p. 83). The design currently most favoured, of which all the above-named FBRs are examples, uses as coolant molten metal, usually sodium. (The Dounreay Fast Reactor uses an alloy of sodium and potassium – usually called 'nak', for Na and K, the chemical symbols of the metals – which is liquid at room temperature.) Such a 'liquid metal fast breeder reactor' (LMFBR) is to be sure not the only possible design; a gas-cooled FBR is also possible and is under development. But liquid metal coolant has obvious advantages. Liquid sodium, being a metal, has a high thermal conductivity; even without moving through an FBR core it can siphon out considerable heat. Furthermore since it boils at the high temperature of 990°C it need not be pressurized, which considerably reduces one major engineering problem.

On the other hand, sodium does have drawbacks. As every boyhood chemist knows, sodium reacts enthusiastically with water; it reacts likewise with a wide range of other materials. Accordingly, although the sodium coolant is not itself pressurized, its open surfaces in an LMFBR circuit are covered by an inert gas such as argon – which in turn tends to get swept into the flowing sodium and cause unwanted bubbles. Unlike gases or water (light or heavy), sodium is opaque, making remote inspection of reactor internals peculiarly difficult. Sodium coolant must not, of course, be allowed to cool below its melting point of 97.5°C anywhere in the circuits, or it solidifies.

Sodium does not readily absorb fast neutrons – if it did it could not be used in a fast-neutron core – but when it does it becomes sodium-24, which is an intensely radioactive gamma emitter. As a result the primary sodium coolant must be confined entirely within the biological shielding of the core. This necessitates a second sodium circuit, with a heat exchanger inside the biological shielding – but itself shielded from neutrons – to pick up the heat from the radioactive primary sodium and carry it out through the shielding to a second heat exchanger in which steam is generated. Steam generators in which molten sodium

Table 1 Typical reactors

Type	Magnox Reactor	Advanced Gas-cooled Reactor (AGR)
Name of example	Dungeness A (UK)	Hinkley Point B (UK)
Heat output	840 MWt	1494 MWt
Electrical output	275 MWe	621 MWe
Efficiency	32.7 %	41.6 %
Fuel	Natural uranium metal rods clad in 'Magnox' alloy	Uranium oxide, 2 % enriched, clad in stainless steel
Weight of fuel	304 tonnes	113.7 tonnes
Fuel burn-up	3850 megawatt-days per tonne	18 000 megawatt-days per tonne
Moderator	Graphite	Graphite
Core dimensions	13.8 metres diameter 7.4 metres high	11 metres diameter 9.8 metres high
Peak power density	1.1 kilowatts per litre	4.5 kilowatts per litre
Coolant	Carbon dioxide gas	Carbon dioxide gas
Coolant pressure	19 atmospheres	40 atmospheres
Coolant outlet temperature	245° C	634° C
Vessel	Welded steel, 0.102 metres thick	Prestressed concrete, 5 metres thick
Refuelling	On load	On load
Comments	Low power density and mass of graphite mean slow temperature rise in fault conditions. Main hazard is low melting point and ignition temperature of Magnox, if air should enter breach in cooling circuit.	Low power density and mass of graphite mean slow temperature rise in fault conditions. Entire primary cooling circuit is enclosed in vessel, and oxide fuel in stainless steel has wide safety margin above operating temperature before melting temperature.

High Temperature Gas-Cooled Reactor (HTGR)	Pressurized Water Reactor (PWR)
Fort St Vrain (USA)	Zion 1 (USA)
842 MWt	3250 MWt
330 MWe	1050 MWe
39.2 %	32.3 %
Uranium carbide particles, 93 % enriched, coated in graphite matrix	Uranium oxide, 3 % enriched, clad in zirconium
16.7 tonnes	99 tonnes
100 000 megawatt-days per tonne	21800 mega watt-days per tonne
Graphite	Water ('light' water)
6 metres diameter	3.35 metres diameter
4.7 metres high	3.6 metres high
6.3 kilowatts per litre	102 kilowatts per litre
Helium gas	Water ('light' water)
46.7 atmospheres	150 atmospheres
785° C	318° C
Prestressed concrete, 4.5 metres thick	Welded steel, 0.203 metres thick
Off load	Off load
Low power density and mass of graphite mean slow temperature rise in fault conditions. Entire primary circuit is enclosed in vessel, and entire core is ceramic, with very high melting temperature. Highly enriched fuel could pose safeguards problem, despite ceramic formulation.	Very high power density. Loss of coolant pressure also involves loss of moderator – shuts down fission reaction but loses heat sink. Fault conditions may produce very rapid temperature rise, possibly even to melting temperature of oxide fuel. Heavy section welded steel pressure vessel requires very high-quality construction, because of very high operating pressure.

Type	Boiling Water Reactor (BWR)	CANDU Reactor
Name of example	Browns Ferry 1 (USA)	Pickering 1 (Canada)
Heat output	3293 MWt	1744 MWt
Electrical output	1065 MWe	508 MWe
Efficiency	32.3%	29.4%
Fuel	Uranium oxide, 2.2% enriched, clad in zircaloy	Natural uranium oxide clad in zircaloy
Weight of fuel	169 tonnes	92.6 tonnes
Fuel burn-up	19 000 megawatt-days per tonne	7000 megawatt-days per tonne
Moderator	Water ('light' water)	Heavy water
Core dimensions	4.8 metres diameter 3.7 metres high	6.4 metres diameter 5.9 metres long
Peak power density	49 kilowatts per litre	16.2 kilowatts per litre
Coolant	Water ('light' water)	Heavy water
Coolant pressure	68 atmospheres	85 atmospheres
Coolant outlet temperature	285°C	293°C
Vessel	Welded steel 0.159 metres thick	Zircaloy pressure tubes 0.1 metres in diameter, 5 millimetres thick
Refuelling	Off load	On load
Comments	High power density. Loss of coolant pressure also involves loss of moderator – shuts down fission reaction but loses heat sink. Fault conditions may produce rapid rise in temperature, possibly even to melting temperature of oxide fuel. Heavy section welded steel vessel requires very high-quality construction because of high operating pressure.	Rather low power density, and cool moderator in separate system, mean slow temperature rise in fault conditions. Pressure tube construction means less likelihood of propagation of a flaw from one tube to others. Fabrication of pressure system involves simpler configurations than those of full-size pressure vessel, despite high coolant pressure.

Steam-Generating Heavy Water Reactor (SGHWR) Winfrith (UK)	Liquid Metal Fast Breeder Reactor (LMFBR) Phénix (France)
309 MWt	563 MWt
94.5 MWe	233 MWe
30.5 %	41.4 %
Uranium oxide, 2.3 % enriched, clad in zirconium	Mixed uranium and plutonium oxides, 20–27 % effective enrichment, clad in stainless steel
21.8 tonnes	4.3 tonnes
21 000 megawatt-days per tonne	100 000 megawatt-days per tonne
Heavy water	None
3.1 metres diameter	1.4 metres diameter
3.66 metres high	0.85 metres high
11.2 kilowatts per litre	646 kilowatts per litre
Water ('light' water)	Liquid sodium
63.5 atmospheres	1 atmosphere
282°C	562°C
Zircaloy pressure tubes 0.13 metres in diameter, 5 millimetres thick	Cylindrical stainless steel pot 12 metres in diameter, 12 metres high
Off load	Off load
Rather low power density, and cool moderator in separate system, mean slow temperature rise in fault conditions. Pressure tube construction means less likelihood of propagation of a flaw from one tube to others. Fabrication of pressure system involves simpler configurations than those of full-size pressure vessel, despite high coolant pressure.	Power density 10–100 times that of 'thermal' reactor designs, but metallic heat conductivity of sodium provides cooling even if circulation fails. System at atmospheric pressure, so no depressurization problem. Fuel is concentrated fissile material – unlike that of 'thermal' reactors. Change of geometry – if coolant flow is interrupted – may produce increase in rate of fission reaction, perhaps very rapid increase. Fuel also presents safeguards problem because of possible misuse of plutonium.

and water are separated only by thin tube walls must be fabricated to very high standards; steam generators have proved to be one of the most troublesome features of LMFBRs.

The world's longest-serving FBR is the British Dounreay Fast Reactor. Its core is only 53 centimetres high, hexagonal, 52 centimetres across each face; a person could easily put his arms around it – although he would be ill-advised to do so. Its maximum power output is 60 MWt, giving 14 MWe. That 60 MWt is generated, note, in a core whose volume is only about 110 litres – a power density of over 500 kilowatts per litre, well over one hundred times that in a Magnox core. The whole reactor is enclosed within a steel containment sphere 41 metres in diameter, of which the lower hemisphere is lined with concrete 1.5 metres thick. The containment and the remote location gave additional safety features during the early development of FBR technology.

The Dounreay Fast Reactor, while supplying a nominal power output, was primarily intended as a laboratory for development of FBR fuel and other technology. It was clear from an early stage that uranium metal fuel, as in the Dounreay Fast Reactor, would not permit operating temperatures high enough to achieve the desired electrical efficiency. Accordingly, the 250-MWe Prototype Fast Reactor (PFR) uses oxide fuel, with its higher melting point. For the PFR the fuel is not just uranium oxide, but a mixture of the oxides of uranium and plutonium. The uranium is only natural uranium; 'depleted' uranium is even better, mixed with enough plutonium to provide the fissile material required. 'Depleted' uranium is uranium from which some of the uranium-235 has been removed to be incorporated in enriched uranium (see pp. 91–5).

The low thermal conductivity of the oxide mixture necessitates making the individual stainless steel fuel pins very narrow to keep the interior temperature from reaching embarrassing heights; a PFR fuel pin is less than 6 millimetres in diameter. There are 4.1 tonnes of mixed oxide fuel in the core, including the equivalent of 1.1 tonnes of plutonium-239 oxide.

The core–blanket array is enclosed in an open-topped tank

full of molten sodium, and sits in a much larger pot of molten sodium. The sodium emerges from the top of the fuel and blanket assemblies and flows through intermediate heat exchangers, giving up its heat to secondary, non-radioactive sodium. Three primary sodium pumps stir the primary potful. The secondary circuits carry the heat out through the shielding to the steam generators.

No pipes or other penetrations enter the primary pot below the level of the sodium, minimizing the possibility of any loss of primary coolant. Above the core, in the reactor roof, is a 'rotating shield' from the bottom of which projects the refuelling assembly.

The PFR, like the other LMFBRs of its generation, is at least partially an experimental facility, aimed at establishing the criteria for a commercial fast reactor. One of these criteria is the rate of breeding achievable. A common measure of this important characteristic of performance is the so-called 'doubling time': the time taken for a breeder reactor to double the amount of fissile material associated with its operation. This inventory of fissile material includes that within the reactor core at a given time, in irradiated fuel elements in the cooling pond, in transit to the reprocessing facility, within the reprocessing facility, in transit to the fuel fabrication plant, within the fuel fabrication plant, in transit back to the reactor and awaiting insertion into the core. This total aggregation of fissile material is the 'pipeline inventory' associated with the FBR. As a rule, in addition to the fissile material within the reactor, there will also be three or four times this amount outside it at other stages of the fuel cycle – perhaps, for a reactor of the size of the PFR, four to five tonnes of plutonium.

The breeding gain is the additional proportion of plutonium contributed during the time a fuel charge spends in the FBR. The smaller this breeding gain, the greater the number of cycles required to double the total amount of plutonium. Accordingly, there are two avenues available through which the doubling time can be shortened: the breeding gain can be increased, or the length of a given fuel cycle can be shortened.

Increasing the breeding gain within the reactor means, essentially, operating at a higher neutron flux; this means decreasing the space between fuel pins, while at the same time requiring a much increased rate of heat removal, criteria clearly in conflict. The only parts of the fuel cycle which can be shortened are those which occur outside the reactor. The obvious step to shorten is the sojourn in the cooling pond. Unfortunately, shortening this sojourn – to as little as thirty days, as has been suggested – means transporting irradiated fuel which is still very radioactive indeed, with all that that entails.

Doubling times in the present generation of FBR cores seem unlikely to be much less than twenty years; some would put the figure much higher. FBR designers are aiming for doubling times of less than ten years; but the engineering required, and the implications for safety of working to such close tolerances, may prove daunting. In the meantime one of the most evident limits on any rapid expansion of construction programmes for FBRs is the availability of plutonium. To get reactor plutonium, it has been suggested that governments may have to cannibalize their arsenals of nuclear weapons. Some would call that the strongest vote in favour of the fast breeders.

3. The Nuclear Fuel Cycle

Undoubtedly the most extraordinary things that happen to reactor fuel happen within the core of an operating reactor. But a great deal happens to it outside the reactor, both before and after its sojourn in the core. The odyssey of the fuel material, from its origin in the earth's crust, takes it from a mine, to a mill, possibly through a specialized facility called an enrichment plant, and through a fuel fabrication plant before it enters the reactor; afterwards it goes to another specialized facility called a fuel reprocessing plant; some of the material thereafter reaches a theoretically final resting place, while the rest may re-enter the process at an earlier stage. The whole succession of processes, with the transport which links them, is called the nuclear fuel cycle. In practice it is not yet very cyclic; but the possibility exists of making it much more so, provided certain problems – both technical and otherwise – can be overcome. Present policies within the nuclear industry are generally directed towards this end. But, cyclic or otherwise, the nuclear fuel cycle outside the reactor gives rise to many of the most controversial aspects of nuclear technology. In the following pages we discuss the fuel cycle, and some of the problems which arise in it.

Uranium Production

Uranium is found in nature as mineralization in sandstones, in quartz pebble conglomerate rocks, and in veins, and to a smaller extent in other types of deposit. There are significant uranium reserves in the US, Canada, southern Africa, Australia, France and elsewhere. High-grade uranium ores contain up to 4 per cent uranium; but known reserves of this quality have been largely worked out, and ore grades ten times lower, 0.4 per cent

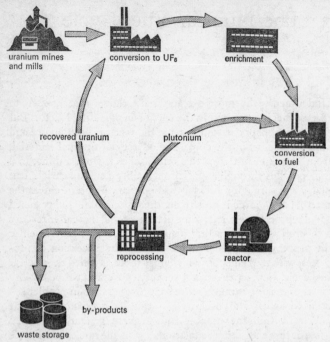

Figure 10   The nuclear fuel cycle

and less, are now being worked. Still lower grades – down to 0.01 per cent and less – are also being noted for development.

Uranium ore deposits are found by a variety of exploratory techniques. The folklore image of the uranium prospector with his Geiger counter, picking his way over the hillside listening for clicks, has little to do with contemporary uranium prospecting. Uranium exploration usually begins in the air, looking for abnormal traces of airborne radioactivity given off by the decay products – so-called 'daughters' – of uranium. Airborne instruments look for tell-tale gamma rays and other evidence of radioactivity. More evidence is assembled on the ground, by studying likely geological formations, by testing samples chemically, and ultimately by drilling.

Uranium ore is extracted by surface or by underground mining. The crude ore is fed into a series of crushing mills, which grind it to the consistency of fine sand. Chemical solvents then dissolve out the uranium, which emerges from the process in the form of a mixture of uranium oxides with a chemical formula equivalent to U_3O_8. This oxide mixture, usually called 'yellow cake', forms the raw material for all the succeeding processes that lead eventually to the reactor core and the chain reaction. Yellow cake contains 85 per cent uranium by weight. Besides the yellow cake, there remains after extraction some one hundred times its weight of residual sand, called 'tailings' – which also contains the radium which had accompanied the uranium. There also remains, per tonne of ore, over 3700 litres of liquid waste, which is both chemically toxic and radioactive. A uranium mine and associated mill may produce over 1000 tonnes of uranium per year, from at least 250 000 tonnes of ore.

Hazards arise at several stages in the uranium production process. The first of these arises from the uranium ore itself, *in situ* and subsequently. When uranium-238 undergoes alpha decay, it produces a succession of further alpha-emitters, including radium-226 and its immediate daughter-product, the chemically inert but radioactive gas radon-222. Any aggregation of uranium which has remained for some time undisturbed – such as a geological deposit – therefore exudes this radioactive gas. When a uranium ore deposit is broken up in mining the escape of the radon is facilitated. Radon-222 is an alpha emitter with a half-life of less than four days, which produces its own radioactive 'daughters'. These radon daughters are however solids. When a radon-222 nucleus in the air emits an alpha particle, the resulting nucleus of polonium-218, being momentarily electrically charged, adheres to any dust particle nearby. Accordingly, air containing radon also contains dust particles laden with intensely radioactive radon daughters. Underground uranium miners who are permitted to inhale such air have proved appallingly susceptible to lung cancer.

The first evidence of this effect was established by 1930, after medical investigations of miners working deposits in

Joachimsthal, in Germany. A similar effect has since appeared in the miners working deposits in the south-western USA after the Second World War. Inadequate ventilation and insufficient expenditure on mine safety have been blamed for the lung cancer deaths of over one hundred American uranium miners; out of a total of some 6000 men who have worked in American underground uranium mines the US Public Health Service has estimated that from 600 to 1100 will die of lung cancer because of radiation exposure on the job.

Uranium mine tailings also present a problem. The military rush for uranium in the USA led to accumulation of vast piles of tailings; estimates range as high as 90 million tonnes, much of this piled on riverbanks in the south-western USA. The consequent radioactive pollution of waterways has represented a serious problem; at one stage inhabitants downstream in the Colorado river basin were exposed, through their drinking water, to three times the ICRP (see Appendix B, p. 281) maximum permissible intake of radium – which is a bone-seeking radionuclide even more dangerous than strontium-90.

Meanwhile it was discovered in the 1960s that the sandy tailings had been used as fill beneath the foundations of many buildings in many communities, notably Grand Junction, Colorado: buildings including homes, schools and hospitals. The radon gas emanating from these building structures into the air indoors now exposes the local inhabitants – including children – to exactly the same radon daughter-products which have already been responsible for thousands of lung cancer deaths of miners from Joachimsthal to Grand Junction. Only in late 1972 did the US government finally agree to contribute funds to an attempt to control the piles of dry tailings still blown freely by the wind across many inhabited areas of the south-western USA. The question of government aid to reconstruct the presently radioactive buildings in Grand Junction and elsewhere remains. The tailings piles will remain dangerously radioactive for tens of thousands of years.

Uranium Enrichment

As indicated in Chapter 2 (pp. 33–4), the fissile uranium-235 nuclei in natural uranium – 7 out of 1000 nuclei – are too dilute to support a chain reaction. Their effectiveness can be increased by interspersing the uranium fuel with a moderator to improve the neutron economy, as already described (pp. 33–4). Alternatively, or in addition, it is possible to increase the proportion of uranium-235 nuclei in the material. This process is called 'uranium enrichment'. Indeed, for weapons applications, it is possible, and often necessary, to have uranium which consists almost wholly of the 235 isotope; such weapons applications use uranium which is at least 90 per cent uranium-235.

Bringing about this increase in the concentration of the 235 isotope is not, however, easy. It cannot be done chemically; in chemistry uranium-235 and uranium-238 are virtually identical. Only their minute difference in mass – 3 units in 235 – can be used as a basis for separation. There are several physical phenomena in which this minute mass difference produces a measurable difference in behaviour between the two isotopes. Of these the phenomenon of earliest large-scale interest was the rate of diffusion through a thin porous membrane. The lighter uranium-235 diffuses just slightly more swiftly through such a membrane; this effect is the basis for what are arguably the largest industrial establishments in the world, the 'gaseous diffusion plants'. There are three such plants in the USA, including the vast Oak Ridge plant which produced the enriched uranium for the Hiroshima bomb; one in Britain, at Capenhurst in Cheshire; one in France, at Pierrelatte – another is under construction; two in the USSR; and at least one in China.

The details of gaseous diffusion technology are – because of their military implications – still to a considerable extent secret. The basis of a gaseous diffusion plant is very simple: a metal-walled cell, with a thin membrane of porous metal dividing it in two. (Fabrication of such membranes, which must withstand lateral pressures and chemical corrosion while providing a selective diffusion barrier, is one area where much detail is still not publicly available.) In order to utilize the different diffusion rates of the two uranium isotopes, it is necessary to convert the original yellow cake, solid uranium oxides, into uranium hexafluoride, UF_6. This compound, called 'hex' for short, is the simplest compound of uranium which can be easily vaporized. Furthermore, fluorine has only one stable isotope; so the different diffusion rates of hex molecules will depend only on the difference between the uranium isotopes involved. It must be added that hex is a viciously corrosive, reactive gas, requiring very careful handling and high-quality metallurgy in the vessels through which it travels.

Under controlled pressure hex enters one chamber of a diffusion cell. It diffuses through the membrane into the other chamber, the lighter-235 isotopes diffusing slightly faster than the heavier-238 isotopes. In a given cell the concentration of 235 can be increased, however, only by about one part in a thousand. Accordingly, the diffusion process must be repeated thousands of times. A cascade arrangement is set up. Gas from the high-pressure chamber of a cell, slightly depleted of the 235 isotope, is piped back to earlier cells; gas from the low-pressure chamber, slightly enriched in the 235 isotope, is piped onwards to later cells. By this means, using thousands of pumps and condensers, it is possible to raise the proportion of 235 isotope to more than 99 per cent. Since pumping heats the gaseous hex the plant must also include large-scale cooling systems.

The uranium whose share of 235 nuclei has been reduced is called, as mentioned earlier, depleted uranium. One factor affecting the performance of a gaseous diffusion plant is the 'tails assay' – the level at which the percentage of 235 is so low that it is no longer worth trying to extract any more 235 from the hex. This tails assay is usually somewhere between 0.2 and 0.3 per cent 235, compared with 0.7 per cent in natural uranium. If the depleted hex is discharged from the plant when it still contains 0.3 per cent 235, more yellow cake will be required to produce a given amount of uranium enriched to a given level; on the other hand if the depleted hex is not discharged until its tails assay is down to 0.2 per cent 235, part of the plant operates with very depleted hex, from which it is even more difficult to extract a useful amount of 235.

In the early stages of enrichment, diffusion cells must be comparatively large; the desirable uranium-235 nuclei are accompanied by a comparatively cumbersome cloud of uranium-238 fellow-travellers. As the proportion of 235 increases, the total mass of hex which must pass through successive cells decreases; the high-enrichment end of the plant uses comparatively small cells, in which only the 235 nuclei remain, with a few stragglers of 238. For this reason the early stages of enrichment, to 3 or 4 per cent uranium-235, require as much pumping as all the stages from this level onwards. The effort expended in the enrichment process is measured in units of 'separative work'; the throughput capacity of the plant is measured in units of separative work per year. Separative work is loosely correlated with the total energy required to carry out an operation – energy to run pumps, et cetera. In general a comparatively large amount of hex enriched to a few per cent requires the same amount of separative work as a comparatively small amount of hex enriched to 90-plus per cent.

All the present generation of gaseous diffusion plants were built under military auspices. Their electrical requirements are awesome; the Oak Ridge plant, in full operation, requires some 2000 megawatts of electricity, enough to power a sizeable city. (Electricity for the Oak Ridge plant is largely provided by fossil-fuel power plants burning strip-mined coal, a nicely ironic touch.) A gaseous diffusion plant likewise takes up an impressive area, as much as half a square kilometre. However, because of the differences between the low-enrichment end of the plant and the high-enrichment end, such plants are not easy to convert from production of strictly military weapons-material involving enrichment to more than 90 per cent uranium-235 to production of fuel for power reactors. Light water reactors, SGHWRs and AGRs require for their fuel a level of enrichment of only 2 to

4 per cent. In consequence, although the first-generation gaseous diffusion plants are in theory available to service present power reactors, other approaches to enrichment are now attracting attention.

France, in partnership with several other countries, is embarked on building a large diffusion plant designed primarily for production of reactor fuel – and therefore quite different in detail from the military plant at Pierrelatte. Some US corporations have been invited by the US government to use hitherto classified enrichment technology in building commercial enrichment plants. But the high initial costs and long construction period are making the private sector very wary of involvement; this may in due course create difficulties. A particularly paradoxical situation has arisen in Quebec province in Canada. Although Canada's CANDU reactors use natural uranium fuel, plans have been proposed to erect a gaseous diffusion plant in the northern wilds of Quebec, as a way to make use of the hydro-electricity generated by the vast James Bay project.

Meanwhile an alternative enrichment technology is making its first contribution to the present-day nuclear fuel cycle. Like the gaseous diffusion process this alternative requires thousands of stages in cascade; the stages this time consist of gas centrifuges. When uranium hexafluoride gas enters a spinning centrifuge, the uranium-238 hex molecules tend to drift to the outer perimeter of the centrifuge chamber, leaving the lighter uranium-235 hex molecules closer to the axis of the chamber. Piping channels the axial hex, slightly enriched, onwards to successive centrifuges, and the perimetral hex, slightly depleted, backwards – just as in the cascades of a diffusion plant. It is claimed that the centrifuge method consumes only one-tenth of the energy required for diffusion, a major advantage for the centrifuge approach. An international cooperative effort between Britain, Germany and the Netherlands is now constructing gas centrifuge enrichment plants at Capenhurst and at Almelo, in the Netherlands.

Other techniques are considered to show promise. One is based on deflection of gas sprayed from a nozzle: the lighter hex-

235 molecules are more easily deflected. But undoubtedly the most exotic technique presently being developed is based on lasers. A laser can be tuned so finely that its radiation ionizes uranium-235 hex molecules while not ionizing uranium-238 hex. It is then necessary somehow to use the handle provided by the electric charge on the ions to sift the 235 hex molecules out of the cloud. Reports indicate that progress is being made. If laser enrichment should prove technically possible, it will undoubtedly be kept secret: unlike the other technologies mentioned, a single stage of laser enrichment could conceivably bring about almost complete separation of non-fissile uranium-238 and fissile uranium-235, offering an alarmingly direct short cut to weapons material, even from uranium ore.

Heavy Water Production

Uranium isotopes are not the only ones requiring separation for nuclear applications. At the other end of the table of elements come the isotopes of hydrogen – of which the second, deuterium, is the best neutron moderator of all. The American and Canadian reactor designs offer in this context a tidy contrast: whereas the Americans enrich the fuel and take the moderator as it comes, the Canadians take the uranium as it comes and, so to speak, enrich the moderator.

The difference in mass between an ordinary hydrogen nucleus and a nucleus of heavy hydrogen or deuterium is proportionally very large; a deuterium nucleus is about twice as massive as a nucleus of ordinary hydrogen. As a result certain types of chemical interchange can be used to separate the light and heavy hydrogen nuclei. The Girdler–Sulphide (GS) process now in large-scale use employs the two chemically similar molecules water and hydrogen sulphide. The former consists of two hydrogen atoms bonded to an oxygen atom, the latter of two hydrogen atoms similarly bonded to a sulphur atom. In a mixture of water molecules and hydrogen sulphide molecules, the distribution of the hydrogen isotopes between the oxygen and sulphur atoms depends on the temperature. At low temperatures – about 25°C – the liquid water contains proportionally more deuterium than it does at higher temperatures – about 100°C. This shift of equilibrium can be used to transfer deuterium atoms out of one batch of water and into another, using hydrogen sulphide as a sort of conveyor belt.

First the water and hydrogen sulphide are mixed together at the lower temperature; deuterium shifts from hydrogen sulphide into water. Some of the enriched water is led off for further enrichment. The rest is fed into a tower at the higher temperature; deuterium now shifts from this water into the hydrogen sulphide. This enriched hydrogen sulphide in turn shuttles back to enrich more water. The depleted water can be discarded, and the enriched water fed onward through a cascade, successively boosting its percentage of deuterium.

Whereas the enrichment of uranium becomes easier the higher it gets, as far as mass transport is concerned, the enrichment of water gradually gets too cumbersome, as the water–hydrogen sulphide exchange reaction becomes inefficient. By this time it is however possible to carry out fractional distillation, utilizing the significantly higher boiling temperature of deuterium oxide – about 101.4°C – to boil away much of the remaining ordinary water. Electrolysis can refine this to a final composition of 99.75 per cent deuterium oxide; by this stage electrolysis is a comparatively inexpensive and efficient way to dispose of the remaining ordinary hydrogen.

There are perhaps a dozen heavy water production plants in all – in the USA, Canada, France, India and elsewhere. The recent upsurge of interest in heavy water reactors – and the high efficiency of such reactors for plutonium production – makes it likely that current production capacities, now of the order of 300 tonnes of deuterium oxide per year per plant, will be hard pressed to keep up with the demand. On the other hand, heavy water is intended to be a permanent part of a reactor system; unlike enriched fuel it is not 'consumed'. Once a reactor is equipped with its operating complement of heavy water its only requirement from then on is enough to replace losses in refuelling and inevitable leakage. Since heavy water now costs upwards of £20 per kilogram operators strive to minimize such losses.

Fuel Fabrication

Fabrication of fuel for reactors is now a major – and complex – industrial process in its own right. In Chapter 3 (pp. 55–7) we encountered some of the determinants affecting reactor fuel and its cladding: ease of heat removal, durability when subject to radiation damage, chemical stability, and physical and mechanical properties which lend themselves to economical fabrication. An additional requirement, at every stage, is establishing and maintaining high purity in the materials, to keep them free of neutron-absorbing impurities. Fuel fabrication facilities accordingly strive to carry out the relevant industrial processes in conditions of cleanliness like those of an operating theatre.

Among the present range of power reactors the only large ones using uranium metal fuel are the British Magnox reactors, and their French cousins. The awkward metallurgy of uranium has already been mentioned. Nonetheless uranium can be fabricated by common metal-working techniques.

The uranium fuel of water-cooled reactors – PWRs, BWRs, CANDUs and SGHWRs – is in the form of uranium dioxide. Uranium dioxide powder is made from uranyl nitrate solution, which may originate either in a uranium mill (from natural uranium), an enrichment plant (from enriched uranium hexafluoride), or reprocessing plant (see pp. 100–103). Fabrication by powder techniques is employed to form the desired shapes – for instance the short cylindrical pellets described in Chapter 3 (pp. 56–7). Baking at high temperature produces stable, dense pellets – the denser the better; high density facilitates the chain reaction, improves the generally poorer thermal conductivity, and also helps to retain the gaseous fission products which accumulate in the fuel material.

T–D

If uranium metallurgy is awkward, plutonium metallurgy is positively fiendish. The metal occurs in six different crystal phases, whose properties change drastically with temperature; two phases even contract, rather than expanding, as temperature increases. Its thermal conductivity is low, its melting point is low, it oxidizes violently on contact with air, and – when its peculiarly vicious radiotoxicity is added to the mix – all in all it is a material without many redeeming virtues. Of course, in the one-shot chain reaction of a fission bomb most of these problems are overcome in a flash. But for the controlled chain reaction in a reactor the choice falls not on the metal but on the dioxide.

Its fabrication is a much more demanding process than that of uranium, requiring much tighter control of quantities and ancillaries. Precautions must prevent not only the escape of the toxic material but also inadvertent juxtaposition in undesirable geometries; unlike most reactor-grade uranium, reactor-grade plutonium is mostly fissile nuclei, and may easily come together in quantity in such a way as to achieve criticality. The consequent barrage of neutrons and gamma rays could cause serious injury or death to anyone nearby. This is particularly hazardous in the case of aqueous solutions of plutonium compounds, since the water acts as a moderator. However, with appropriate precautions plutonium dioxide can be processed like uranium dioxide; indeed, the two oxides, mixed in suitable proportions, are a particularly promising form of fuel material.

Fuel rods or fuel pellets, once fabricated, are clad as described in Chapter 3 (pp. 56–7), and where appropriate arranged in assemblies for transport to the reactor.

Transport

One of the major advantages claimed for central electricity generation with a nuclear heat source is the relatively small bulk and mass of fuel and waste that must be transported to and from the station. A fossil-fuel station requires so much coal or oil that it is economically advisable to situate the station near the fuel supply. A nuclear power station, on the other hand, requires at most one or two shipments a week; furthermore the shipments away from the station are far more massive and bulky than those to the station. Fresh fuel elements, only minimally radioactive, can be and are shipped in ordinary cases like any other cargo. But once irradiated they must be heavily shielded, so that a shipment of two tonnes of irradiated fuel requires a fifty-tonne steel shipping cask.

Fresh reactor fuel and fuel materials are shipped by rail, by road, by water and by air, in increasing quantities every year, between various parts of the fuel cycle. Apart from the usual protections against low-level radioactivity around the shipping cases, the main technical consideration is to guard against stacking of cases so close together that the aggregation of fissile material can reach criticality (but see pp. 246–53). Elaborate codes of practice are published to provide appropriate technical guidelines.

The shipping of irradiated fuel is something else again. The irradiated fuel must be handled remotely at every stage, from loading at the station – usually from a cooling pond – to unloading at the reprocessing plant (see pp. 100–103). For short journeys irradiated fuel elements usually travel in massive water-filled casks which are vaned on the exterior to help dissipate the decay heat. For longer journeys, especially those by sea, the casks must be coupled to cooling circuitry.

An accident involving a shipment of irradiated fuel could release dangerous amounts of radioactivity. This is particularly the case if – as is now being mooted – the fuel is shipped after not 120 days, or even 90 days, in a cooling pond but after as little as 30 days, when even short-lived radioisotopes still contribute significantly to the radioactivity. Such short cooling periods are economically desirable; reactor fuel is expensive, and leaving it out of circulation in a cooling pond does not help to balance the books. On the other hand, shipping arrangements for such fiercely radioactive used fuel will be themselves more costly. As it is, shipping casks must pass severe tests, typically a thirty-minute fire after a ten-metre fall. But as the number of shipments increases so does the likelihood of an accident.

Fuel Reprocessing

One feature distinguishes nuclear power technology from all others: the left-overs. Unlike the ash, say, from a coal-fired power station, the used fuel from a nuclear power station contains both very valuable material and uniquely troublesome waste. Recall that the first large reactors were built expressly so that, under neutron bombardment, the uranium-238 in the fuel would be transmuted into plutonium-239. This plutonium had to be recovered from the fuel, as did the unused uranium-235 which was left after poisoning of the chain reaction by fission products and other effects had made it necessary to remove the fuel. The same requirements hold today. Fissile uranium and plutonium are much too valuable to throw away. Even if it were not valuable, plutonium is in any case far too dangerous to be let loose in the environment (see pp. 235–41 and Appendix B, p. 285). Nor must the remains of the fuel, including the fission products, be thrown away – not because of their value but because they too are dangerously radioactive. Accordingly, the irradiated fuel from a reactor is usually 'reprocessed'.

A nuclear fuel reprocessing plant is a chemical plant – but no ordinary chemical plant. Because its raw material, irradiated reactor fuel, is intensely radioactive, all the operations must be carried out by remote control, behind heavy shielding. The process equipment must be highly reliable, and require a minimum of maintenance. Once in operation it is contaminated by the radioactivity, and any malfunction necessitates months, or indeed years, of decontamination before it can be set right. Accordingly, the process line uses as few mechanical parts as possible, and depends instead on gravity flow and simple valves.

Different designs of fuel require different handling. The British reprocessing plant at Windscale was originally set up to process metal fuel elements from plutonium-production and Magnox reactors. Magnox fuel is stored in a cooling pond adjacent to the reprocessing plant. When ready for reprocessing it is transferred under water, by operators watching on closed-circuit television, into the building and up into the first of a series of 'caves' or 'hot cells'. The walls of the caves are of concrete some two metres thick, to intercept the gamma radiation from the fission products in the fuel. Once in the caves the fuel can be observed through special windows built into the cave walls. Each window is like a large aquarium, filled with a solution of a chemical such as zinc bromide, which is virtually transparent to visible light but strongly absorbs the very short wavelengths of gamma radiation.

A Magnox fuel element entering the reprocessing caves is picked up by remote control and dropped on a stripping machine which chops off the ends of the element and unzips the Magnox cladding as easily as peeling a banana. The contaminated cladding drops on to a conveyor belt to be transported to another building nearby, which looks like an aircraft hangar but is actually a heavy concrete storage bin. The bare fuel rod is loaded into a transfer magazine and thence dropped into a vat of nitric acid, which dissolves it ready for reprocessing.

Oxide fuel must be treated slightly differently. When the Windscale plant was modernized in the early 1960s the old reprocessing plant was converted into a 'head end' plant to prepare oxide fuel for reprocessing. During their sojourn in the reactor the pellets of oxide fuel swell, and wedge themselves inextricably inside their tubular cladding. Accordingly, no attempt is made to strip off the metal cladding from the oxide pellets. The irradiated fuel, in a transport cask, is lifted to the topmost storey of the plant. The cask is clamped to the end of a line of caves, and the fuel element fed gradually into the first cave. In this cave an awesome ram-powered shear chops through the entire element, which may consist of more than 100 fuel pins. Each chop produces a burst of pulverized pellets and a barrage of centimetre-long rings of cladding, which drop into nitric acid. The acid dissolves the remnants of the pellets. The rings of cladding are left behind, to be stored, like the Magnox cladding, indefinitely. The acid stream passes on to the main reprocessing plant.

The reprocessing of fuel for the high temperature gas-cooled reactor (HTGR) will require a very different form of head end. One possibility is to grind up the ceramic fuel and burn off the carbon in an oxygen-fed burner before dissolving the remainder. One design of HTGR fuel uses two kinds of fuel particles, with the fertile thorium particles coated in a silicon layer. The separation process can then keep these particles, and their newly-bred fissile uranium-233, from being mixed with other uranium during reprocessing.

From the head end onwards reprocessing involves liquids: the dissolved fuel, and a succession of various solvents. First the nitric acid solution is mixed with a solution of an organic solvent; in the Windscale plant this solvent has a polysyllabic name which is unceremoniously abbreviated to TBP/OK. The uranium and plutonium cross over into the TBP/OK, leaving behind about 99.96 per cent of the fission products in the water-based acid. This acid stream carries these fission products out of the reprocessing plant; its subsequent progress will be described in the following section.

Almost all the uranium and plutonium (not quite all – see p. 111) are now in the TBP/OK stream, flowing under gravity from one section of the plant to the next. After another pass through a similar 'solvent extraction' stage, to remove lingering fission products, this stream now encounters another water-based solution; the plutonium and uranium at this point part company, the plutonium returning to the water solution, leaving the uranium behind in the TBP/OK. Various chemical man-oeuvres occur en route, to bring about these shifts of loyalty. Eventually, after repetitions of some steps, the uranium emerges in one stream and the plutonium in another, as uranyl nitrate solution and plutonium nitrate solution respectively, ready to return to an earlier point of the fuel cycle and be made into fresh fuel elements (or weapons). Process streams of concentrated fissile material – especially plutonium – must be designed to guard against accidental criticality. Reprocessing plants have elaborate alarm systems to warn personnel in the event of a criticality accident – which may of course be quite invisible, despite the fusillade of neutrons and gamma rays.

Radioactive Waste Management

Throughout the nuclear fuel cycle, the materials involved share one common property: they are all to some extent radioactive. Radioactivity arises in mining and milling; the materials remain radioactive throughout enrichment, fuel fabrication and transport – but their activity is not particularly intense. This, however, changes dramatically once they find themselves inside an operating reactor: neutrons from reactor cores tend to make their entire neighbourhood radioactive. So long as the materials in this neighbourhood remain within the biological shielding all is well; but radioactivity inevitably finds a number of escape routes from the confines of reactors, however well buttoned up. The most important of these is via refuelling, when the entire radioactive inventory of the fuel is taken for reprocessing; we shall discuss the consequences of reprocessing in a moment. But radioactivity also makes its way directly out of an operating reactor, and must be dealt with.

Any radioactivity which emerges into the environment outside the biological shield in the course of routine reactor operation is called a 'running release' or 'routine release'. The simplest kind of running release originates just inside the biological shield itself. In reactors with concrete shielding close to the core, it is desirable to keep the concrete from being exposed directly to the heat of the core. Accordingly, except in the case of pre-stressed concrete pressure vessels, a thin layer of air is blown up the inside wall of the concrete and discharged to the atmosphere from a stack atop the reactor building. Some of the nuclei in the air absorb neutrons and become radioactive, a process called 'neutron activation'. The most notable of the 'activation products' is argon-41, a radioisotope of the inert gas argon. Some reactors are known to discharge hundreds of thousands of curies of argon-41 annually. Fortunately, however, argon-41 has a very short half-life, only some 1.8 hours: so this apparently enormous output decays to a very low activity before drifting from the stack down to ground level. Other atoms in the air also become activated, but only in small amounts and/or for very short half-lives.

Reactor coolant may carry radioactivity out of the biological shield. Impurities in water or graphite moderator are susceptible to neutron activation. Carbon – from graphite moderator or carbon dioxide coolant or both – can become radioactive carbon-14. But since normal carbon is carbon-12 the transmutation requires the absorption of not one but two neutrons and is accordingly infrequent. Heavy water coolant can absorb neutrons, turning the deuterium (hydrogen-2) into hydrogen-3, or tritium, which is radioactive. But the one coolant which does respond readily to neutron activation is the sodium coolant in liquid metal fast breeder reactors. As already indicated it becomes sodium-24, so intensely gamma-active that it must be kept entirely within the biological shield.

The fuel cladding too may contribute to the activity in the cooling circuit, as the cladding suffers gradual corrosion by the hot coolant. Again it is primarily a consequence of impurities in the cladding – which has, of course, been made as little susceptible to neutron absorption as possible, for reasons of neutron economy. The worst offenders in this category are impurities in zircaloy cladding on water-cooled reactor fuel. Corrosion of this cladding is enhanced by the intimate contact with the fast-moving fluid at high pressure, which quickly carries surface corrosion into the coolant flow.

Much more serious are leaks from the fuel cladding, to which some reactors seem prone. The build-up of gaseous fission products inside a fuel rod imposes increasing strain on the cladding; if for any reason the cladding develops a flaw the fission gas seeks it out and escapes into the coolant. A more sizeable leak also releases the volatile fission products, among them the dangerous iodine-131 (see pp. 142, 147–8, 164–5).

'Burst can detection gear' in a Magnox reactor sniffs out tell-tale radioactivity in the coolant, and locates faulty fuel elements. If the reactor can be refuelled on load, it is possible to remove leaking fuel without a shutdown. If the reactor, like most water-cooled designs, can only be refuelled off load, a shutdown would be necessary. Furthermore fuel which is leaking slowly may be difficult to locate. In any case, early replacement of fuel disrupts the fuel cycle, and distorts the planned pattern of neutron density in the core. For all these reasons leaking fuel is frequently left in a reactor until routine refuelling.

These effects make it necessary to decontaminate the cooling circuits of a reactor. Otherwise the unavoidable leakage of radioactivity through valve seals and other permeable points becomes a potential hazard to personnel, and may interfere with maintenance. Decontamination is usually done routinely, by bleeding off a small portion of the coolant and replacing it with fresh uncontaminated coolant. Of course, each time a refuelling machine is coupled to a reactor vessel for on-load refuelling the machine acquires a share of the activity in the coolant; this activity must be discharged – and kept track of – when the machine is depressurized. In the case of carbon dioxide coolant the gas bled off – the 'off gas' – is passed through a variety of filters and delay stages.

Boiling water reactors, which pass the primary coolant directly through turbines, are especially prone to leakage of active coolant. One possible procedure in such a case is to provide storage tanks for coolant bled off. In such tanks the coolant activity can be allowed to decay for some months before it is released. Similar hold-up tanks may also be provided for drainage from floors, for water from sinks used in decontamination, and for water from the laundry in which contaminated clothing is cleaned. The cooling ponds for storing irradiated fuel prior to shipment for reprocessing usually pick up activity from the exterior of the fuel elements and also from any leaking elements; so this cooling water, too, has to be dealt with. A common procedure is to cycle the water slowly through the ponds, continually diverting a small fraction, mixing and diluting it with the much greater mass of cooling water discharged from the turbine condensers back into the river or coastal water from which it has been abstracted. Water treatment systems may also use ion exchangers and other standard forms of purification installations to collect and segregate radioactive compounds in sludges.

In the course of everyday business in a reactor plant a certain amount of solid material also becomes contaminated with radioactivity – floor mops, paper towels, broken glassware from sampling labs, et cetera. The volume of contaminated solids ought to be reduced by, say, incineration; but at present they are usually simply buried in designated burial grounds, or dumped at sea in prepared containers, as are active ion-exchange sludges.

All of the radioactivity which reaches the outside world directly from a reactor installation by these routes may be lumped together as 'low-level' radioactivity. Because running releases are as a rule dilute and not very radioactive it is often said that reactors discharge very little radioactivity to their surroundings. This is, strictly speaking, true; but it is slightly misleading. Almost all the radioactivity which leaves a reactor does so within the used fuel removed from its core. Since the reactor has created virtually the whole of this radioactivity, it is somewhat special pleading to imply that the radioactivity thenceforth bears no relation to the reactor. On the contrary: the radioactivity from used reactor fuel is one of the most challenging problems posed by operation of nuclear reactors.

When a fresh fuel element enters a reactor it is as sleek and glossy as a surgical implant. When it emerges again after irradiation it is swollen, discoloured, possibly even caked with what the nuclear engineers bluntly call 'crud'. Inside the beat-up cladding the fuel now contains unused uranium-235 and -238, plus a wide assortment of other nuclei created by the fission reactions, neutron absorption and radioactive decay: uranium-237, plutonium-239, -240, -241 and -242, americium-241 and other so-called 'actinides', and literally hundreds of different fission-product nuclei and their decay and neutron-activation products, including krypton-85, strontium-89 and

-90, iodine-129 and -131, and caesium-137. Some of these species have short half-lives; while the irradiated fuel sits in the cooling pond or travels from the reactor to the reprocessing plant the short-lived isotopes like uranium-237 and iodine-131 decay to insignificance. By the time the fuel enters the threshold of the reprocessing plant after, say, 100 days of cooling, its radioactivity arises mainly from radioisotopes of only about a dozen elements.

Assuming that the cladding has been gas-tight until the head end plant strips or chops it open, the gaseous fission products, particularly krypton-85, thereupon emerge into the atmosphere of the hot cell. Krypton-85 has a half-life of about 10.8 years. It is a chemically inactive inert gas, and is accordingly difficult to recapture. Present practice is simply to discharge it from a stack into the outside air. (Until 1971 American authorities regarded the amount of krypton-85 discharged from their reprocessing plants as classified – that is, secret – because it might reveal how much fissile material they had produced.) Radiobiologists do not consider that the consequent gradual build-up of gamma-emitting krypton in the global atmosphere offers any present hazards. But if nuclear power programmes expand as widely as anticipated it is assumed that some form of krypton retention will have to be installed in reprocessing plants before the turn of the century. Some similar considerations also apply to iodine-129. It has a half-life of 16 million years, and is therefore not very radioactive; but since it is concentrated, like all isotopes of iodine, in the human thyroid any significant build-up in the environment must be viewed with caution.

Within the reprocessing plant there is also an inevitable accumulation of low-level liquid and solid radioactive wastes exactly like those which collect in a reactor plant – indeed probably more copious. Solid wastes are buried, or dumped at sea, as before. At the Windscale reprocessing plant low-level liquid wastes are discharged into the Solway Firth through twin pipelines emptying under water more than three kilometres offshore, at the rate of some 500 000 litres per day.

All such routine releases of radioactivity are carried out in

accordance with standards laid down by appropriate national
authorities, usually based on the guidelines of the International
Commission on Radiological Protection (see Appendix B,
pp. 281–3). In Britain, for instance, discharges of radioactive
effluent are monitored by the Ministry of Agriculture, Fisheries
and Food, as well as by the dischargers themselves, to ensure
compliance with standards imposed as conditions of operation
in licences issued by, among others, the Inspectorate of Nuclear
Installations, under national legislation. In the USA, as we shall
describe in Chapter 5 and Appendix B, such standards have
been the subject of protracted controversy.

The fuel apart from the gaseous or volatile fission products
and the cladding – and in some reprocessing plants including
the cladding – is dissolved in the nitric acid for first-stage
separation. The witches' brew left behind when the uranium
and plutonium pass over into the organic solvent is called 'high-
level waste'. Without doubt it is the most daunting waste
material produced in any industrial process. At Windscale, the
reprocessing of one tonne of fuel produces about five cubic
metres of high-level waste – that is, about enough to fill five or
six bathtubs. The waste contains nitric acid, fission products
which are both thermally hot and intensely radioactive, iron
from corrosion of plant vessels, chemical impurities from the
original fuel, and a dash of carried-over organic solvent. As
may be imagined, it requires delicate treatment, to avoid
unpleasant side-effects like reactions between solvent and nitric
acid at high temperatures. The volume is reduced by evaporation – under vacuum, to keep down temperatures. The procedure must be carried out under careful control – always
remotely of course – to avoid crystallization or precipitation
where it might prove embarrassing (such as in process lines)
and to keep the fission products at a low concentration so that
the rate of heat output does not overwhelm the cooling system.

After evaporation the concentrated waste is led to the storage
facility near the main reprocessing plant. At Windscale this is a
concrete building containing an array of special storage tanks,
eight each of 70 cubic metres capacity and three – so far – of

150 cubic metres capacity. The smaller tanks are fitted with cooling coils; each of the larger tanks has seven independent cooling circuits, external water jackets which include leak detectors, and an internal system of agitators to prevent solids from settling onto the bottom of the tank. The cooling circuits on one of the larger tanks can remove up to 2 megawatts of heat – that is, about 13 watts per litre; this in turn limits the permissible concentration of the fission-product stream flowing into the tank. The temperature in the tank is kept to about 50°C. Gradual evaporation of the water from the solution is accompanied by gradual decrease of the radioactive heat output; evaporation can be kept in step with heat output to maintain sufficiently low concentration. It is also necessary to prevent a build-up of hydrogen, produced by the breakdown of water molecules by radiation – so-called 'radiolytic hydrogen'. Tanks can be interconnected, to prevent overloading of cooling circuits with incoming waste of comparatively high output, and to provide transfer facilities in case of a leak. Tanks in use are permanently sealed into massive steel-lined concrete shielding vaults, never to be seen again. A programme of construction of new tanks keeps spare capacity available. At the end of 1974 the total volume of liquid waste stored at Windscale was about 600 cubic metres.

Similar tank storage installations are located at reprocessing facilities in the USA, France, Belgium, the Soviet Union, India, China and elsewhere. The most famous – or notorious – is at the Hanford Reservation in Washington state. Here, in 151 very large tanks, is stored the high-level liquid waste – nearly 250 000 cubic metres – resulting from recovery of the plutonium from the Hanford production reactors, for the US nuclear weapons programme. It is generally reckoned that tank storage of high level wastes can only allow for a useful life of twenty to twenty-five years per tank, albeit perhaps somewhat longer for tanks of stainless steel. Many of the Hanford tanks are ordinary carbon steel; more than a dozen leaks have already occurred, including at least one very large leak indeed.

On 20 April 1973 nobody paid any particular attention to tank 106T in the 200 West area of the Hanford Reservation. The T tank farm was built in 1943–4 and is one of the oldest at Hanford. It includes twelve tanks each with a capacity of just over 2 million litres, and four with a capacity of just over 200 000 litres. Tank 106T is one of the larger ones. It is made of reinforced concrete with a carbon steel liner on its bottom and sides, cylindrical in shape, about 23 metres in diameter and 10 metres deep, sunk in the ground with about 2 metres of earth over its domed top. In April 1973 tank 106T contained high-level radioactive waste from the Purex fuel reprocessing plant – about 1.5 million litres of it, mostly liquid. On or about 20 April tank 106T sprang a leak.

The employees of the Atlantic Richfield Hanford Company (ARHCO), AEC contractors responsible for the facility, went about their business. From 4 to 24 April a fresh instalment of hot liquid was pumped into tank 106T. Every week or so after pumping ended someone read the gauge indicating the level of the radioactive liquid in tank 106T, and jotted it down in a log. Every week the day shift supervisor left it to someone else to review the data. Apparently nobody did. On 8 May the monthly reading of radioactivity in Well 299-W-10-51, a test hole next to tank 106T, showed nothing unusual. On 31 May, when next the radiation detector was lowered into the well, it went off scale. The monitoring operator told the day shift supervisor, who asked him to check this well daily, but did nothing further. The next day, using a less sensitive Geiger-Mueller probe the monitoring operator got a reading of 300 000 counts per minute. The radiation data was placed on the desk of the supervisor, who did not review it. This went on daily until, in a leisurely fashion, by means of a casual exchange of telephone calls on 7 June, the idea occurred to someone that perhaps all was not entirely in order under tank 106T. On the morning of 8 June the supervisor confirmed that there did indeed seem to be a leak in the tank. Plans – 'emergency' plans, curiously out of keeping with the foregoing proceedings – were initiated to pump out tank 106T. That afternoon the AEC was notified.

Between 20 April and 8 June tank 106T leaked approximately 435 000 litres of highly radioactive liquid into the earth, containing *approximately 40 000 curies of caesium-137, 14 000 curies of strontium-90, and 4 curies of plutonium. AEC investigators later declared that the radioactivity would not reach the level of the ground water below the tanks; but a drilling programme to locate the lost waste had to be curtailed, lest new drill-holes facilitate the downward migration of radioactivity.*

ARHCO personnel told the investigation committee that they had not been able to find a method by which they could assure them-selves that a tank was free of leaks before refilling it. On 24 June 1973 ARHCO informed the AEC that a further extension of deadline would be needed for starting the formal quality assurance programme which had already been set back before. The leak was the eleventh recorded at Hanford; it was not the last.

Clearly the hazardous life-span of some constituents of high-level waste far outreaches that of a storage tank. Strontium-90 has a half-life of 28 years, caesium-137 one of 30 years. It takes ten half-lives to reduce the radioactivity of a sample a thousandfold ($\frac{1}{2} \times \frac{1}{2} \times \frac{1}{2} \times \frac{1}{2} \times \frac{1}{2} \times \frac{1}{2} \times \frac{1}{2} \times \frac{1}{2} \times \frac{1}{2} \times \frac{1}{2}$ equals $1/1024$). Accordingly, it takes about 300 years for the radioactivity of 1 curie of strontium-90 or caesium-137 to drop to 1 millicurie. The high-level waste from 1 tonne of irradiated fuel includes about 100 000 curies of each. A 1000 MWe PWR produces at least 25 tonnes of irradiated fuel per year – that is, well over 2 million curies of strontium-90 and another 2 million curies of caesium-137. Some 300 years hence this particular contribution will have dwindled by a factor of a thousand, to only 2000 curies of each: except that 2000 curies of strontium-90 is not very 'only'. Multiply such figures by the number of reactors now in operation, under construction or planned, and the magnitude of the consequent problem becomes numbingly apparent. Furthermore, present methods do not – for economic reasons, if not for technical reasons – extract all the actinides from the fission product waste: perhaps 1 per cent of the plutonium, with its 24 400-year half-life, is left behind to add to the unpleasantness of the residue.

It is evident that such quantities of potentially dangerous

radioactivity require scrupulous stewardship. While tank storage is regarded as a satisfactory interim measure, efforts continue to devise a long-term solution to the problem. (After all,
the cost of perpetual replacement of tanks is an *outré* item in a
balance sheet – although it appears as just that.) Some of the
suggestions seem outlandish. It has been seriously proposed
that high-level waste might be fired from the earth by rocket into
the sun; but the cost would be – excuse the expression –
astronomical, requiring eventually several Saturn rockets per
week, with the ever-present possibility of a rocket failure
scattering its load of waste over the earth. Other futuristic
proposals include placing canisters of waste on the Antarctic
ice cap, where their own heat would let them melt their way
down to bedrock; and dumping canisters of waste onto locations
on the ocean floor where geological movement might gradually
swallow them into the earth's interior. But a more realistic view
is that we are stuck with however much high-level waste we
create – and so are our children, and our children's children, for
centuries – indeed millennia – hence.

To mitigate this burden somewhat, it is now accepted that
high-level waste ought at least to be in solid rather than in
liquid form, to immobilize the waste, reducing the possibility
of it spreading via leaks or vaporization. Several approaches for
solidification are under development. In the USA, at Hanford,
wastes are simply being allowed to boil themselves dry inside
storage tanks, to be left as solid cake in the tanks. At the National Reactor Testing Station, in Idaho, high-level waste is
'calcined' – baked at high temperature – into granules like
coarse white sand, which are stored in huge concrete-shielded
stainless steel bins underground. Another approach, favoured
particularly by the British, is to evaporate and fuse the high-
level waste into dense glass. Plans are now in preparation to
build a pilot scheme which will imbed high-level liquid waste
in hollow glass pillars about 75 centimetres in diameter
and 4 metres tall. The surface temperature of such a pillar
would be 450°C; clearly it will still need some form of cooling.
The plan is to store the pillars in a cooling pond; in order to

eliminate the possibility of waste leaking out of the glass pillar it will be formed inside a container which can be sealed.

One question of interest is whether such 'vitrification' of waste should be carried out directly as the waste leaves the reprocessing plant, or only after, say, three years of storage as a liquid. If the waste is left in the liquid state for three years, radioactive decay will reduce its heat output per unit volume; it will then be possible to incorporate more waste in a given volume of glass without encountering heat-removal problems after vitrification. On the other hand, if the waste is vitrified only after interim storage, there will always be on hand a substantial inventory of the most highly active waste in liquid form, which tends to offset the safety advantage of immobilizing the waste.

While disposal into the sun or under an icecap still seems far-fetched, one form of permanent disposal has moved beyond the purely conceptual: insertion of canisters of solidified high-level waste into a stable geological formation. The type of geological formulation which is considered most suitable is the salt dome, and investigations in the USA and West Germany have suggested several possible locations. A hole would be drilled into the floor of an underground gallery in a salt dome. A waste canister would be lowered into the hole – by remote control, as usual – and loose salt would be poured in after it. The salt would become soft under the intense heat from the canister, and would snuggle close around it, sealing it permanently in place and conducting heat away at an adequate rate to keep the solid waste from melting. 'Salt Vault', a pilot scheme for salt-formation storage, was carried out in the USA in the late 1960s near Lyons, Kansas. But – despite earlier official pronouncements to the contrary – the site was eventually abandoned as unsuitable. A company mining salt on a nearby location pumped several thousand cubic metres of water down a drill-hole to bring up dissolved salt; but the water disappeared, casting doubt on the alleged impermeability of the salt formation. Fortunately no high-level waste had yet been buried in it. The search for more reliable salt formations continues. In the USA the location of current interest is Carlsbad Caverns in New Mexico; in West Germany storage of low-activity waste is already taking place in the salt caverns of Asse, with plans to commence storage of higher-activity waste shortly.

At least one nation with major nuclear commitments has for the present refused explicitly to consider any form of 'non-retrievable' disposal for high-level waste. Canada has announced that all high-level waste arising in her nuclear power programme will be stored above ground in a retrievable form until any alternative approach has been fully proven.

It is worth noting in passing that the nuclear industry refers, as a matter of course, to 'waste management'. It looks like a career with a future – a long future.

Part Two

The World and Nuclear Fission

4. Beginnings

Purely as a physical phenomenon nuclear fission offers ample scope for intellectual problem-solving. If it implied nothing further it could be left to those specialists who might find satisfaction in its intellectual challenge; the rest of us could busy ourselves with other more pressing concerns. Unfortunately, nuclear fission – as everyone knows – implies much more than abstruse mathematical argument and donnish hair-splitting. Almost from the time it was first recognized, in the late 1930s, nuclear fission has implied not merely articles in learned journals but major decisions of public policy. The social, economic and political context of nuclear fission has been from the beginning an essential factor in its development; in turn, it has exerted an extraordinary range of social, economic and political influence. To foresee with any clarity the shape of the nuclear future, a historical perspective is imperative. It is necessary to know not only how nuclear fission occurs, but also who makes it occur, under what circumstances and for what purposes. We have already alluded, in a preliminary way, to these aspects of the subject. It is now time to step back and examine them in much more detail. Two themes in particular emerged early on: the unforeseen effects of radiation, and the undisclosed results of nuclear weapons development. As we shall see, nuclear activities from the outset have been characterized by unpredictability and secrecy. Throughout nuclear history, either too little has been known, or enough has been known but too little said.

In 1896 Henri Becquerel discovered radioactivity. Shortly thereafter, by carrying a vial of radium in his pocket and burning himself, he also discovered the most troublesome attribute of radioactivity: its biological effects, actual and potential. An episode which occurred more than fifty years later characterizes the situation that has prevailed since Becquerel's discoveries. Interested organizations were debating the design of an international symbol to convey the warning 'DANGER: RADIATION'. One group of participants, including labour union representatives, favoured the design of a grinning skull with an aura of wavy lines emanating from it. But spokesmen for government and industry groups flatly refused to sanction such a design, which they considered too frightening. As a result the design finally adopted was a circle with three leaves fanning out from it – intelligible only to those to whom it has been explained beforehand, and utterly devoid of prior associations, either benevolent or malevolent.

This contretemps is a succinct instance of the schism which divides viewpoints about radiation. As indicated earlier the very biological essentials of the issue are hotly controversial, increasingly so; but we shall confine discussion of them mainly to Appendix B (p. 285ff). What is in some ways yet more controversial is the developing social context of radiation, especially that produced by radioactivity. (A good case can be made for concern about other forms of ionizing radiation, especially diagnostic X-rays; but we shall here comment only that diagnostic X-rays should be used only when clearly indicated by medical evidence, and should be generated only by well-shielded apparatus for as brief an exposure as possible.) Before plunging into the tumult of nuclear controversy it is important to stress – since it may later be easy to overlook – that the main health problem is that created by radiation; that the radiation is inherently invisible and detectable only by special instruments; that different forms of radiation present different hazards (see Appendix B, p. 285ff); and that the deleterious result of exposure to radiation may not manifest itself for many years. For these reasons it is far from easy to be confident of understanding the effects of radiation. Accordingly, human undertakings involving radioactivity may be peculiarly difficult to evaluate from a public-health standpoint. They may also, as we shall see, be difficult to evaluate by a variety of other criteria, not least the economic.

After Becquerel's discovery came those by Pierre and Marie Curie, who isolated from the uranium ore pitchblende the powerfully radioactive elements polonium and radium. It now appears that the Curies were, paradoxically, fortunate in their poverty. Their laboratory was a draughty attic, and its otherwise undesirable ventilation probably saved Marie Curie from an early death brought on by inhalation of radon from her experimental materials. (The draughty attic did not save her husband, who died of quite another technology, under the wheels of a cart in a Paris street.)

Scientific fascination with the newly discovered radioactive substances was almost at once paralleled by a search for practical applications. Roentgen's X-rays were, within months of their discovery, applied in medicine; but within three years the X-rays – which of course required apparatus to generate them – were meeting competition from the radiation of radium and its relatives. Alas for the early successes of radiotherapy: its pioneers, and Marie Curie herself, were among the first to experience the insidious delayed consequences of radiation. So were their patients, some of whom died not from cancer or other afflictions but from radiation burns inflicted with the aim of healing. Radium became for a time a fashionable material. 'Radium spas' were suddenly successful in several parts of Europe, and doctors prescribed medicines containing radium. There was also a vogue for 'luminous dial' wrist-watches: the digits on the watch faces were painted over with a mixture of zinc sulphide and radium, and glowed in the dark. The women who worked in the watch factory used fine brushes to apply this luminous paint; and to give a brush a suitably pointed tip its user would lick it. In due course almost all the 'luminizers' fell sick, with bleeding gums and anaemia, and eventually most developed bone sarcoma – cancer of the bone – from the accumulation of radium in their bodies. A small New Jersey plant alone produced more than forty victims from staff employed between 1915 and 1926.

The radium which now seemed so ubiquitous was being produced from mines like the old silver mine of Joachimsthal, whose uranium ore gave it a new lease on life. But the miners of Joachimsthal – as we have described – were prone to *Bergkrankheit*: lung cancer, induced by inhalation of radon and radon daughters. Medical detective work had by 1930 identified the genesis of this disease, and made clear that it could be prevented only by ensuring thorough ventilation of underground uranium mines, a lesson which was to be callously ignored in the American uranium rush of the 1950s. There was, throughout the 1920s, a growing awareness among medical researchers, biologists, and radiation workers themselves, that radiation had some unpleasant attributes. But there was, throughout this time, no particular public concern, and no general sense of controversy about radioactive materials and their uses. It was as if the controversy over radiation had an even longer latency period than radiogenic disease.

From 1939 to 1945 there was neither opportunity nor inclination to question the circumstances which might arise in the manufacture and use of radioactive materials: no opportunity, because most of the frantic effort then occurring was under strict conditions of secrecy, and no inclination because those involved were preoccupied with the much more immediate and pressing fear that Nazi Germany would achieve nuclear weapons-technology first. As all the world now knows, the Nazis did not. The USA achieved the technology and the weapons, and used them, bringing an abrupt and devastating end to the Second World War.

In the aftermath of the Hiroshima and Nagasaki bombs, the US government set up the Atomic Bomb Casualty Commission. Its function was dual. It was – so far as the Japanese victims were concerned – a source of medical aid for those who had survived the nuclear explosions but had already suffered or might subsequently suffer from the effects of the radiation. It was also – so far as the USA was concerned – an agency which could carry out a large-scale study and documentation of the effects of radiation on human beings. As may be apparent the two roles were not wholly compatible. Many Japanese grew to resent deeply the role which they felt they were being forced to play, as guinea-pigs for the further enlightenment of the world's first and only user of nuclear weapons.

Nonetheless the US medical researchers studying radiation effects also found opportunities closer to home. Fifteen days after the Hiroshima bomb, on 21 August 1945, Harry Daghlian, a physicist at Los Alamos, accidentally allowed a sample of fissile material to reach criticality while he was handling it. His hands and body were raked by a massive burst of radiation, gamma rays and neutrons. Admitted to hospital within half an hour, Daghlian lost sensation in his fingers, then complained of internal pains and finally became delirious. His hair fell out. His white blood cell count surged as his shattered tissues tried vainly to cope. It took him twenty-four days to die.

Daghlian's death brought home to the entire Los Alamos community the grim ethical conflict in which nuclear physicists – the 'atomic scientists' – now found themselves. It is worth stressing, more than thirty years later, that the first to recognize the dilemma of nuclear energy – the conflict between its constructive and destructive potentials – were the nuclear physicists themselves.

Even before the dropping of the Hiroshima bomb a group of those who had helped to create it signed a memorandum subsequently known as the Franck report, submitted to the US Secretary of War on 11 June 1945, forecasting with dismaying accuracy the nuclear arms race, in the event of the use of the bomb against a military target. James Franck and his colleagues proposed instead that the bomb be demonstrated in a remote site, desert or island, before representatives of Japan and of the allied United Nations, and that the USA then renounce its use thenceforth, provided that the rest of the nations of the world agreed to do likewise. But the Franck proposals, as the world knows, fell on stoney ground – unlike the Hiroshima and Nagasaki bombs. The initiators of this report, with colleagues similarly concerned, founded, later in 1945, the *Bulletin of the Atomic Scientists*. Since its inception the *Bulletin*, published in Chicago, has remained one of the most perceptive and committed voices addressing nuclear controversy of every kind (see Bibliography, p. 295, for further information).

On 21 May 1946 Louis Slotin, a Canadian physicist working at Los Alamos, was performing an exercise he called 'twisting the dragon's tail'. He had done the operation many times, while determining experimentally the details of fast critical assembly of the hemispheres of uranium-235 which had to slam together to produce the desired nuclear explosion. In this way Slotin had determined experimentally the critical mass for the Hiroshima bomb. Slotin's exercise involved sliding the two hemispheres gradually towards one another along a rod, using two screwdrivers, and watching the neutron detectors display the build-up to criticality. On 21 May he was showing the phenomenon to a group of a half-dozen colleagues, when the screwdriver slipped. The room filled with blue light. Slotin tore the hemispheres apart with his bare hands, and in so doing probably saved the lives of his colleagues. But he himself was doomed, and knew it. He died nine days later, his terminal radiation sickness being meticulously charted by otherwise helpless medical staff. Slotin's Los Alamos colleagues were forbidden for security reasons to alter their daily routine or reveal anything about the accident, while they witnessed his lingering death.

The assembly that killed Louis Slotin was earmarked for the second bomb in a pair of weapons-tests which the US Navy was preparing to carry out at Bikini atoll in the Marshall Islands. The Marshall Islands had been before the Second World War a German protectorate; this responsibility was taken over by the USA at the end of the war, but the 'protection' thereafter given sounds reminiscent of the 'protection' referred to by racketeers. Tests 'Able' and 'Baker' were carried out 30 June and 25 July 1946. 'Able' was an atmospheric nuclear explosion, 'Baker' one deep underwater in Bikini's lagoon. Both drew bitter protests from many scientists, notably the Federation of Atomic Scientists, a newly-formed federation of local groups of scientists concerned about the implications of their work; subsequently it became the Federation of American Scientists, and thirty years later is still deeply involved in and outspoken about nuclear issues. It was said that the US Navy staged the Bikini tests primarily to show that the Army was not the only branch of the military with nuclear capability. About 42 000 onlookers arrived in 250 ships and 150 planes – military, media, politicians, diplomats, plus additional thousands of scientists with a piece of the action. Scientists not thus involved insisted that the tests would serve little genuine experimental purpose, that they would appear as mere sabre-rattling while the UN grappled with the problem of international control of nuclear technology. To them and many others the whole business looked like a grisly public relations exercise to show off the USA's latest accomplishment.

Whatever their *raison d'être*, the Bikini tests had an aspect which the public did not learn at the time. In order to clear the necessary room for their activities, the US Navy in March 1946 unceremoniously evicted 167 Marshall Islanders from Bikini, transporting them to Rongerik, another island many kilometres distant, with much poorer vegetation, soil and fishing, making expansive promises to the evicted islanders which were thereafter quietly forgotten. The Navy repeatedly assured the islanders that all would soon be able to return to their homes. What they did not add was that the 1946 tests played havoc with the fertile lagoon of Bikini, leaving it full of radioactive mud and making marine life for more than 150 kilometres unsafe to eat. Not until 1968 were the first nine islanders permitted to return to Bikini, to an island altered almost beyond recognition by the nuclear explosions and their after-effects. The mangled ruins of US weapons sheds and towers loomed out of the overgrowth. The new vegetation was coarse and unproductive; even the coconut crabs, huge tree-climbing crustaceans looked upon as a Marshall Island delicacy, had accumulated so much strontium-90 in their shells that the islanders had to be forbidden to eat them. The first Bikini tests were called, grandiloquently, Operation Crossroads. For the Marshall Islanders they must have looked more like a dead end.

No sooner had the Hiroshima and Nagasaki bombs brought a stunning conclusion to the Second World War than the first manifestation of nuclear paranoia became manifest. The bombs had of course been made through a combined effort of US, British and Canadian scientists and engineers; but the Manhattan Engineer District – code name given to the bomb-development project – had all its major facilities in the USA. By mid summer 1945 the Americans had virtually taken over the project, including its direction and, more importantly, the information it had generated. Only days after the Nagasaki bomb a bill was presented to Congress whose ultimate effect, as the McMahon Act of 1946, was to make it illegal for Americans thenceforth to give their erstwhile allies any further access to information about nuclear energy. Top-level discussions, including those between the three heads of government, were contradictory in content and inconclusive in outcome. Eventually the three wartime partners embarked on separate programmes. The evanescent hopes for effective international control of nuclear energy were stillborn. Instead, with the American weapons tests at the Pacific islands of Bikini, beginning in 1946, and the first Soviet atomic bomb test in 1949, the nuclear arms race began. It has seemed ever since to be a race that no one will win.

The McMahon Act – the Atomic Energy Act 1946, to give it its proper title – was not at the time regarded even by British and Canadian scientists – much less their American colleagues – as a particularly unfortunate step. The scientists, it is true, realized too late the Act's divisive implications. But it was initially hailed as a victory for rationality, in that it specifically overrode any military claim to control of nuclear developments. Instead, the McMahon Act established two civilian bodies to exercise this responsibility and control: the United States Atomic Energy Commission (AEC), and the Congressional Joint Committee on Atomic Energy (JCAE). The Act gave the AEC complete control over the funding and direction of postwar nuclear research and development, military and otherwise. The JCAE was to be the Congressional watchdog over the

AEC, the channel through which the elected representatives of the public would monitor and oversee American activities in the nuclear field.

The McMahon Act brought an immediate order into the post-Nagasaki nuclear situation, within the USA at any rate. On 1 January 1947 the AEC came formally into being. It qualified for a healthy slice of the Federal budget, and took over all the facilities established for the Manhattan Project; from that date onwards the American nuclear effort assumed a new dimension.

The AEC's fundamental responsibility was of course development of more powerful and efficient nuclear weapons, and provision of the infrastructure to build them in quantity. Undoubtedly the fiercest controversy in the early years of the AEC centred on whether or not to pursue development of a new form of nuclear weapon, long referred to simply as the 'Super'. There is a limit to the amount of fissile material that can be slammed together efficiently into a prompt critical configuration. Accordingly there is a limit to the energy release possible in a pure fission bomb. Since this energy release is equivalent to that of several hundred thousand tonnes of TNT – several hundred 'kilotons' – it might be thought sufficient for most purposes; but Soviet and American weapons designers thought otherwise.

At least some of them did. Robert Oppenheimer, the brilliant wartime director of the Los Alamos Laboratory, thought the Super was ill-advised, and made no secret of his opinion; in due course this was made the basis of one of the shoddiest episodes in American scientific history, the 'trial' of Oppenheimer in April 1954 which permanently deprived him of access to the nuclear information he above all had been instrumental in developing.

The Oppenheimer case underlined an AEC attitude which was to persist long after AEC interests had diversified into civilian applications of nuclear energy. One of the extraordinary powers granted by the McMahon Act permitted the AEC to call upon the services of branches of the Executive –

such as the FBI and the CIA. The AEC spent millions every year on exhaustive vetting of employees, nominally for reasons of 'national security'; and the AEC's control of access to information established a pattern that was subsequently difficult to break, even in contexts to all appearances non-military.

The principle of the Super was straightforward. If a fission bomb is surrounded with material containing nuclei of heavy hydrogen – deuterium or, better still, tritium (hydrogen-3, with one proton and two neutrons in its nucleus) – the ferocious heat of the fission blast speeds up the light nuclei so that they collide and stick together as helium nuclei. Each 'fusion' of two hydrogen nuclei into a helium nucleus releases a burst of neutrons and additional nuclear energy – once again, mass is converted into energy.

Since there is no immediate upper limit on the amount of 'fusible' material that can be so triggered, the energy release of a fission–fusion or 'thermonuclear' bomb – more commonly known as a hydrogen bomb – is effectively unlimited. Further improvements – if that is the correct word – are also possible. The fusion reaction, like the fission reaction, releases free fast neutrons. Accordingly, a triple-decker bomb can be made: a fission bomb surrounded by fusible hydrogen surrounded by ordinary uranium – much cheaper than heavy hydrogen, and unlimited by criticality considerations. The outer layer of uranium intercepts the barrage of neutrons from within it and undergoes fission, adding yet more energy to the total – and, incidentally, adding also an enormous additional contribution of fission products, far more than those from the small fission 'trigger'.

Through the late 1940s and into the early 1950s one overriding concern dominated nuclear controversy: the accelerating arms race between the USA and the USSR, and the pursuit of the fusion weapon. Espionage, counter-espionage and the Cold War climate made nuclear secrecy and nuclear secrets a fountainhead of collective paranoia. Whatever the effect of the

McMahon Act within the USA, the erstwhile partners of the USA took it as an act of betrayal. The fury and resentment it engendered still linger in the upper echelons of the British and Canadian nuclear communities. Even the official US announcement of the Hiroshima bomb was regarded in Britain as taking too much of the credit (if such it was) for the USA; on 6 August 1945 the British Prime Minister issued a stiff statement pointing out the key roles played by the British and Canadian contributors to the Manhattan project. However, once the initial grievance had been at least suppressed, the British and Canadian governments reacted quite differently. The Canadians decided that Canada neither wanted nor could build nuclear weapons. The British scientists who had been engaged at the Montreal laboratory of the wartime project were recalled to Britain – indeed the Canadians felt somewhat as though, having constructed substantial installations, they were being left holding an expensive and useless bag. We shall return to the Canadian effort shortly.

In Britain, the McMahon Act rankled deeply, and it was never seriously doubted that Britain must thereupon embark on her own nuclear-weapons programme. The British public – and that includes almost all of Parliament – knew, it must be said, virtually nothing of this. Only a passing reference on 12 May 1948, in a House of Commons answer by the Minister of Defence, gave any indication of the furious activity then underway: 'Research and development continue to receive the highest priority in the defence field, and all types of weapons, including atomic weapons, are being developed.' That was all; the Minister would not elaborate, since it was 'not in the public interest' to do so. The organization given the task of developing British nuclear weapons was the Division of Atomic Energy Production, Ministry of Supply, eventually to become the United Kingdom Atomic Energy Authority.

In only two and a half years the Production Division had completed the Springfields uranium and fuel-fabrication plant; the first Windscale pile loaded with Springfields fuel went critical in July 1950. At this time the construction of the reprocessing plant had not even begun; the first irradiated fuel entered the reprocessing plant in late February 1952. On 3 October 1952 the first British nuclear bomb vaporized the frigate *Plym* in the waters of the Monte Bello Islands just off the north-western coast of Australia.

Several nations besides Britain, the USA and Canada had an early foothold in nuclear matters. Germany, Poland, Hungary and other eastern European countries were the origins of many of the scientists who took their abilities to Britain and the USA after the advent of the Nazis – and, to be sure, of some who did not. French scientists participated in the wartime deliberations that led to the Manhattan project. Norway had the Vemork heavy water production plant – until it was blown up by Norwegian partisans in 1943. The USSR was keenly interested in nuclear matters well before the Second World War.

Of these nations the first to embark on a serious nuclear research and development programme was the USSR. Like every other nation then and since, the USSR regarded nuclear matters as government matters, not to be left to industry or academia. In 1943, even as the invading Germany army was well inside Soviet borders, the Soviet government set up a nuclear weapons research institute in Moscow, directed by Igor Kurchatov, and later to bear his name. The Soviet nuclear programme was fully as intense as that of the Americans, leading to a fission bomb in August 1949, and a thermonuclear bomb four years later to the month.

In 1950, when President Truman gave the go-ahead for US development of the Super, another huge AEC facility was established: the Savannah River complex in South Carolina, with more plutonium production reactors, this time moderated by heavy water, a full-fledged reprocessing plant, waste storage, the lot. But the Super remained elusive. The Americans are commonly credited with having detonated the first thermonuclear explosion on Eniwetok, in the Marshall Islands, on 1

November 1952; but it was in no sense an 'H-bomb'. It was an explosion of a large-scale experimental installation, nearly sixty tonnes of delicate equipment; it could no more be dropped on an enemy than could an entire factory. The Soviet thermo-nuclear explosion of 12 August 1953 was a true H-bomb, portable and droppable. At least a sizeable part of the AEC had other things on their minds when on 8 December 1953 President Eisenhower delivered a major address to the UN proposing a programme of 'Atoms for Peace'. On 1 March 1954 the US detonated its first H-bomb, designated Castle Bravo. They expected an explosion equivalent to 7 million tonnes – megatons – of TNT. They got one equivalent to 15 megatons (see pp. 133–5).

The President's speech nonetheless presaged yet a further transformation in the American nuclear scene. A new Atomic Energy Act in 1954 made it possible for private contractors to build reactors and to possess fissile material under licence from the AEC, and declassified a variety of useful information. However the short-term prospects for nuclear electric power generation did not look very enticing to US industry. The technology was certainly promising, and reasonably hard-headed estimates of probable generating costs would have been persuasive, but for the prevailing low costs of oil, and – even more competitive – the surging abundance of indigenous natural gas and coal.

There was little doubt that power reactors of several designs could be built, and that they would indeed generate electricity at a cost which was not unreasonable. But only the most sanguine anticipated that power reactors could compete economically in the USA with fossil-fuel generating units before the mid 1960s. On the other hand, it was clear that if this new technology were to establish itself it could not be expected to be economically viable from birth. The AEC accordingly began to draw up plans for a Cooperative Power Reactor Development Project, and US industry began to show tentative signs of interest.

The US superiority in nuclear experience and resources gave

T–E

them an embarrassment of choices when it came to civil applications; but that same wealth of resources also made the economic context unpropitious. The British situation was precisely the reverse. Their straitened circumstances meant that they had to focus their efforts within a very narrow range of technology. But the context of post-war European economics, with limited supplies of other fuels available, at moderately stiff prices, meant that power production by nuclear means looked appetizing. By the late 1940s the British Atomic Energy Research Establishment at Harwell was attacking with gusto the design problems posed by power reactors.

In August 1952 the military Chiefs of Staff issued a call for greatly-increased production of weapons plutonium. A power reactor producing plutonium as a by-product would for the nonce have to be regarded as a plutonium reactor producing power as a by-product. Designs for such a dual-purpose reactor had already been undertaken, labelled 'Pippa' ('pressurized pile producing power and plutonium'). Apart from the transposition of the last two items, 'Pippa' was, in March 1953, given the go-ahead. We know it now as the first Calder Hall reactor. In some respects its design was Hobson's choice – neither sufficient enriched uranium nor sufficient heavy water could be guaranteed for otherwise more desirable designs, and the fast breeder did not come into the category of production reactor for short-term purposes. But a long and thoughtful study of the economic status of a natural uranium power reactor was already extant – it had been prepared by R. V. Moore of Harwell in the early autumn of 1950, comparing the performance of a 90-MWe nuclear station with that of a similar coal-fired station (see pp. 218–21). It identified the essential basis of this economic comparison, which was to become a perennial: coal involves low capital costs and high running costs, whereas nuclear power involves high capital costs and low running costs. Moore's analysis showed that the crossover point, at which nuclear electricity from natural uranium became cheaper per unit than coal-fired, was well within attainable nuclear criteria.

In the 1970s, some of Moore's assumptions are poignant to recall – an interest rate of 4 per cent, for instance. But the essence of his argument has remained remarkably rélevant. Since the world price of oil began its spectacular climb, the attractions of nuclear power generation have been strongly reinforced. The point at which it becomes cheaper, including both capital charges and running costs, to generate a unit of electricity from uranium rather than from oil, has shifted markedly in favour of uranium. Environmental and manpower considerations in the coal industry have produced a similar shift in the balance between uranium and coal. Of course such questions of cost are not the whole nuclear story – nor are the questions of cost themselves so readily answered, as we shall see in Chapter 8.

5. Out of the Background

Meanwhile the weapons tests continued, and set the stage for the first major public involvement in nuclear policy. The issue was to centre on the radioactive waste from nuclear explosions: the 'fallout'. After the US tests in the Pacific in 1946 and 1948, the AEC established in 1951 a more convenient location at Mercury, Nevada. The Nevada Test Site was originally intended for smaller weapons, but the size criterion gradually dwindled in importance, and an increasing proportion of US tests were carried out at Nevada until the Partial Test Ban of 1963 brought them home one and all. Even in 1951 the explosions at the Nevada Test Site caused problems. Radioactive debris deposited in the mid-western USA contaminated straw used for packing photographic film, and Eastman Kodak, contemplating their ruined film, were irate. The AEC agreed to notify Kodak of fallout patterns that might affect film packaging. The AEC proved to be somewhat less attentive to the possibility that fallout might also affect other features of the surroundings, like human beings.

Yucca Flats was the locale for the Upshot–Knothole series of twelve nuclear tests in 1953, some of which were carried out with soldiers and observers within three kilometres of the blast. On 25 April 1953 a 43-kiloton test at Yucca Flats code-named 'Simon' won itself a measure of distinction. 'Simon' was an atmospheric test, exploded less than 100 metres above the desert. Its radioactive cloud surged to a height of more than ten kilometres; the upper portion then began to drift northeast. On 27 April, about 200 kilometres north of New York City, the radioactive cloud encountered a violent thunderstorm. That morning Professor Herbert Clark of Rensselaer Polytechnic Institute was preparing to carry out some experiments with his radiochemistry students. To their surprise they found that all their Geiger counters were giving readings much higher than normal. Readings outdoors proved even higher, especially near the rainwater. Professor Clark called a colleague in the AEC Health and Safety Laboratory in New York City, and a major survey was set in motion. The public subsequently learned only that the 'rainout' had deposited from 35 to 70 curies of radioactivity per square kilometre around Troy and Albany. The detailed AEC report of the survey was, however, classified. Professor Clark and his students went on to note similar surges of radioactivity in the water reservoirs serving Troy and Albany after a number of later tests in the Upshot–Knothole series. The Troy experience was to provide the first input to an on-going accumulation of data about public exposure to nuclear radiation at low levels.

Meanwhile, the citizens of St George, Utah, Las Vegas, Nevada and other centres much closer to the Nevada site than New York State, were finding the test explosions uncomfortably near, as one after another dumped radioactivity on them. The AEC was unconcerned. In its thirteenth semi-annual report in 1953 it noted that 'Fallout radioactivity is far below the level which could cause a detectable increase in mutations, or inheritable variations'; and that the only possible hazard to humans from animals fed on grass contaminated by strontium-90 would arise from 'the ingestion of bone splinters which might be intermingled with muscle tissue during butchering and cutting of the meat'.

In any case the people of Troy and St George were fortunate, compared to 236 inhabitants of Rongelaap, Rongerik, and Uterik in the Marshall Islands, and twenty-three Japanese crew members of a fishing vessel called the *Fukuryu Maru*. As we have already noted, the USA detonated its first true hydrogen bomb, code-named 'Castle Bravo', on 1 March 1954, in the Bikini atoll. Its yield had been expected to be about seven megatons; it proved to be more than double this. An American destroyer found itself in the path of the radioactive dust; its crew responded by carrying out radiation drill, battening hatches, stationing all hands below decks and waiting until fixed hoses had cleansed the contamination off the exterior surfaces of the ship. But no one had told the islanders or the Japanese fishermen anything about radiation drill.

Uterik, Rongerik and Rongelaap are about 160 kilometres east of Bikini. But wind, in a direction not anticipated by the weapons testers, carried the bomb debris all the way to the other three islands. On 11 March the AEC issued a press statement:

During the course of a routine atomic test in the Marshall Islands twenty-eight United States personnel and 236 residents were transported from neighbouring atolls to Kwajalein island according to plan as a precautionary measure. These individuals were unexpectedly exposed to some radioactivity. There were no burns. All were reported well. After the completion of the atomic tests, the natives will be returned to their homes.

Roger Rapoport, a perceptive American reporter, has noted drily (see Bibliography, p. 292), that the evacuations were indeed according to 'plan', but that the plan was not devised until after the accident occurred:

The victims sustained beta burns, spotty epilations of the head, skin lesions, pigment changes and scarring. And many of the natives did not feel well at all. They suffered from anorexia (appetite depression), nausea, vomiting and transient depression of the formed elements in their blood. Over the next sixteen years twenty-one of the natives on Rongelaap island would develop thyroid abnormalities and thyroidectomies would be conducted on eighteen of them. All but two of the nineteen children who were less than ten years old when the accident happened developed thyroid abnormalities; and two of them were dwarfed for life.

For three years after Castle Bravo Rongelaap remained too radioactive for the islanders to return. In 1972 Lekoj Anjain, one of the Rongelaap islanders caught by the 'Bravo' fallout, died of leukaemia, at the age of 19.

The intervention of the US Navy helicopter service to Kwajalein undoubtedly served a double purpose so far as the stricken Marshall Islanders were concerned. It brought them much needed medical attention; it also ensured that this attention was not accompanied by other attention potentially more embarrassing to the weapons testers. The US Navy did not know, however, that its patrol aircraft had overlooked a Japanese fishing boat, the *Fukuryu Maru*. The *Fukuryu Maru* was trawling for tuna east of Bikini on 1 March, beyond the perimeter of the delineated test zone, when it suddenly seemed to the fishermen as though the sun were rising in the west. Within a few hours the boat was dusted with white ash, sifting down on to the decks, and over the hair and clothing of the crew. By evening two of the crew were vomiting, and overcome by dizziness. By 3 March others were suffering similar symptoms, with aching eyes and itching skin. There was clearly something wrong. The fishing boat turned and made for its home port of Yaizu. It arrived a fortnight later, with all hands suffering from radiation sickness, and the boat still contaminated with radioactivity. Six months later some of the crew were still in hospital. On 23 September 1954 the radio operator, Aiticki Kuboyama, died.

The fate of the *Fukuryu Maru* made headlines all over the world. The irony of its name – translated, it means *Lucky Dragon* – gave an additional twist to the grim saga. The Japanese had been the first victims of the atomic bomb; now it could be said that a Japanese was the first to die from the effect of a hydrogen bomb. The traumatic jolt of the *Fukuryu Maru* incident reinforced the profound psychological revulsion with which the Japanese regarded nuclear energy. Three decades after the Hiroshima and Nagasaki bombs, and two decades after the *Fukuryu Maru*, the Japanese distrust of nuclear energy remains as deep-seated as ever.

The radioactive return of the *Fukuryu Maru* alerted the world, with stunning impact, to the phenomenon of 'radioactive fallout'. Only a month later, in April 1954, India called for a standstill on nuclear weapons tests; needless to say the call drew little response (see Figure 11).

The *Fukuryu Maru* incident was followed by the US Atomic

Atmospheric Explosions

	USA	USSR	UK	France	China
1945	3				
1946	1				
1947					
1948	3				
1949		1			
1950					
1951	15	2			
1952	10		1		
1953	11	2	2		
1954	6	2			
1955	31	4			
1956	14	7	6		
1957	26	13	7		
1958	53	26	5		
1959					
1960				3	
1961		30		1	
1962	38	41			
1963					
1964					1
1965					1
1966				5	3
1967				3	2
1968				5	1
1969					1
1970				8	1
1971				5	1
1972				3	2
1973				5	1

Figure 11 Atmospheric and underground nuclear explosions 1945–73

Underground and underwater explosions

	USA	USSR	UK	France	China
1945					
1946	1				
1947					
1948					
1949					
1950					
1951	1				
1952					
1953					
1954					
1955	2				
1956					
1957	2				
1958	13				
1959					
1960					
1961	9	2		1	
1962	50	1	2	1	
1963	25			3	
1964	28	6	1	3	
1965	28	9	1	4	
1966	40	14		1	
1967	28	14			
1968	37	12			
1969	28	15			1
1970	30	13			
1971	11	18			
1972	7	21			
1973	9	14			

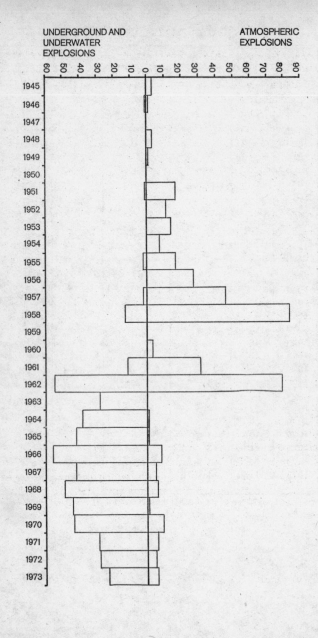

UNDERGROUND AND
UNDERWATER
EXPLOSIONS

ATMOSPHERIC
EXPLOSIONS

60 50 40 30 20 10 0 10 20 30 40 50 60 70 80 90

1945
1946
1947
1948
1949
1950
1951
1952
1953
1954
1955
1956
1957
1958
1959
1960
1961
1962
1963
1964
1965
1966
1967
1968
1969
1970
1971
1972
1973

Energy Act 1954, which declassified some – albeit by no means all – data on nuclear weapons tests. Scientists who examined the data newly available, and compared it with the 1953 AEC comments on possible fallout hazards, found alarming discrepancies. One of the first and most effective critics of the AEC's performance was the nuclear physicist Ralph Lapp. In an incisive series of articles in the *Bulletin of the Atomic Scientists*, beginning in November 1954, he pointed out some home truths the AEC were reluctant to acknowledge. The AEC had assumed, erroneously, that fallout would be deposited more or less uniformly over the entire globe. It turned out, however, to be confined largely to the hemisphere of origin, in the middle latitudes – that is, over the most populous areas of the earth. Furthermore much of it came down within months, rather than remaining safely aloft for years while isotopes of short half-life decayed. In any case, it was pointed out sternly, what mattered was not some putative average radiation dose from fallout, but actual peak doses from local concentrations, like that which arose after the 1953 rainout over Troy. (The significance of the Troy rainout was not in fact identified until Lapp drew attention to it in an article in *Science* in 1962.)

In 1955, the UN General Assembly adopted a resolution establishing a scientific committee to inquire into the effects of radiation and nuclear tests, the UN Scientific Committee on the Effects of Atomic Radiation (UNSCEAR). But the jockeying between the nuclear-weapons powers continued unabated, as did the tests and the fallout, and the efforts of scientists and the public to bring some sense into the proceedings. Bertrand Russell drafted an appeal, co-signed by Albert Einstein two days before his death, which became known as

Figure 11 Atmospheric and underground nuclear explosions 1945–73. Bars to the right of the line represent atmospheric explosions, bars to the left of the line underground and underwater explosions; the 1963 Partial Test Ban Treaty, signed by the USA, USSR and UK, banned atmospheric tests. The 23 US and 33 USSR explosions whose exact dates are unknown are excluded. Source: *World Armaments and Disarmament*

Table 2 Nuclear explosions, 1945–73

After the two US tests at Bikini in 1946 the momentum of nuclear testing built up rapidly. The USA, USSR and UK observed a voluntary moratorium from autumn 1958 to autumn 1961, during which France commenced testing. On 5 August 1963 the Partial Test Ban Treaty came into effect; although tests by the USA, USSR, and UK from that date onward have been carried out underground, testing has continued at a steady rate unimpeded by the ban, with France – and from 1964 China – testing in the atmosphere.

	USA	USSR	UK	France	China	Total
1945	3					3
1946	2					2
1947	0					0
1948	3					3
1949	0	1				1
1950	0	0				0
1951	16	2				18
1952	10	0	1			11
1953	11	2	2			15
1954	6	2	0			8
1955	15	4	0			19
1956	14	7	6			27
1957	28	13	7			48
1958	66	26	5			97
1959	0	0	0			0
1960	0	0	0	3		3
1961	9	32	0	2		43
1962	88	42	2	1		133
1963 (up to 5 August)	11	0	0	2		13

the Russell–Einstein Manifesto. It called upon scientists of all nations to unite to seek a way out of the impasse to which nuclear discoveries had brought mankind. Professor Joseph Rotblat, who had left Los Alamos when it became clear that Nazi Germany could not manufacture nuclear weapons, undertook the organization of such an unofficial scientific conference. An American millionaire agreed to play host to the first gathering, at his birthplace in Pugwash, Nova Scotia. The Pugwash movement grew from this first meeting into one of

	USA	USSR	UK	France	China	Total
1963 (after 5 August)	14	0	0	1		15
1964	28	6	1	3	1	39
1965	28	9	1	4	1	43
1966	40	14	0	6	3	63
1967	28	14	0	3	2	47
1968	37	12	0	5	1	55
1969	28	15	0	0	2	45
1970	30	13	0	8	1	53
1971	11	18	0	5	1	35
1972	7	21	0	3	2	33
1973*	9	14	0	5	1	29
Total	565†	300‡	25	51	15	956

* Figures for 1973 are preliminary.

† Including 23 US explosions conducted between 15 September 1961 and 20 August 1963. Their dates are not specified in the lists available; at least one of them must have been conducted after 5 August 1963.

‡ Including 33 USSR explosions of unknown date up to 1958

---

the most effective and influential international avenues of contact between leading scientists of the USA, the USSR and other countries; its central objective was – and remains – to devise ways to control nuclear developments, to reduce the unparalleled threats they entail.

By 1956 the AEC was willing to concede that people were more likely to drink milk than to consume bone splinters. In its earlier pronouncements it had somehow overlooked the fact that if an animal – say a cow – eats grass sprinkled with strontium-90, chemically similar to calcium, not only its bones but also its milk contains the radioisotope. The AEC belatedly agreed that milk was much the most significant source of strontium-90 in the human diet. In 1956 Norman Cousins, editor of the influential *Saturday Review*, founded with many colleagues the National Committee for a Sane Nuclear Policy, usually just called SANE. Adlai Stevenson, Democratic Presidential candidate, called for an agreement to stop nuclear testing; his Vice-Presidential opponent Richard Nixon denounced Stevenson's proposal as 'catastrophic nonsense'. Nobel Prize-winning chemist Linus Pauling drew up a petition – eventually signed by 11 021 scientists in forty-eight countries – demanding a ban on nuclear tests.

In 1957 the AEC Biological and Medical Advisory Committee concluded, contrary to earlier AEC assertions, that fallout from nuclear tests through 1956, far from being devoid of genetic significance, already seemed likely to produce between 2500 and 13 000 major genetic defects per year in the global population. Meanwhile scientists outside the AEC were pinpointing troublesome radioisotopes which the AEC had either failed to take seriously or ignored completely – radioisotopes like carbon-14, fingered by Pauling, and iodine-131, by California geneticist E. B. Lewis. AEC fallout surveys, it became evident, usually missed iodine-131 completely, because of its short eight-day half-life – despite the fact that iodine-131, as Lewis stressed, would be concentrated like all iodine isotopes in the human thyroid, and might therefore do damage – especially to children – far out of proportion to its concentration in the external surroundings. The inconsistencies in official pronouncements were becoming too blatant to stomach; from May through July 1957 the Congressional Joint Committee on Atomic Energy (JCAE) held hearings into 'The Nature of Radioactive Fallout and its Effects on Man'. However, AEC witnesses insisted that the tests were safe, and the Committee accepted the AEC's protestations, despite some startling exchanges in cross-examination which would suggest that the AEC was wilfully misrepresenting what it knew only too well (see Metzger in Bibliography, p. 292, for examples).

In 1957, largely as a result of pressure from the public and the scientific community outside the AEC, the US Public Health Service instituted a fallout monitoring system, which soon expanded into a widespread network taking frequent samples; before long these fallout monitors had identified strontium-89 and -90, iodine-131 and other hazardous radioisotopes in quantities copious enough to be far from reassuring.

Ralph Lapp's classic study *The Voyage of the Lucky Dragon* (Bibliography, p. 290) demonstrated that the AEC was anything but scrupulous in its stewardship of the public welfare, and the agitation grew. So, relentlessly, did the test programmes. Lest matters seem too cosily domestic to Americans, the British weighed in with their first hydrogen bomb explosion at Christmas Island, on 15 May 1957.

The British hydrogen bomb of May 1957 seems to have spurred on the USA and the USSR to redoubled efforts of their own. In the USA, of course, Americans had been subjected to their own fallout since the commencement of testing at the Nevada site in 1951; the Windscale fire (see pp. 162–6), whose escaping radioactivity crossed the Channel and settled in some parts of northern Europe, gave Europeans their first encounter with indigenous fallout of a sort, albeit a comparatively decorous encounter. Neither the Americans nor the Europeans were exactly happy with the state of affairs; and early 1958 saw the birth of two organizations which were going to play distinctive roles in nuclear policy in the coming years. Despite the similarity of their initials – CND in the UK, and CNI in the US – their stances were strikingly different, and offer an opportunity to contrast two characteristic categories of public response to nuclear issues.

CND, the Campaign for Nuclear Disarmament, grew out of a movement founded in north London in the mid 1950s, through an influential *New Statesman* article by J. B. Priestley, to a meeting at Central Hall Westminster in February 1958 which drew 2000 people, at which the movement became CND. Behind the scenes a more action-oriented group centred on *Peace News* put forward a plan for a march from London to the Atomic Weapons Research Establishment at Aldermaston, to take place on Easter week-end. The response to the plan dumbfounded the organizers. Five thousand people assembled in Trafalgar Square on Good Friday to set out for Aldermaston. With CND leading the way through miserable weather they walked for four days; their ranks swelled to 10 000 for the culminating rally outside the barbed wire of the Aldermaston establishment. The slogan of the Aldermaston March was 'Ban the Bomb'. The CND symbol, the semaphore signals for ND – nuclear disarmament – three arms, at four, eight, and twelve o'clock inside a circle, was eventually to become a universal 'peace' sign, later to appear on the helmets of soldiers in Vietnam and other equally incongruous places.

It would be an odds-on wager that few people who now sport the symbol know its origin, or indeed regard the aim of 'nuclear disarmament' as of other than historical interest. But in the heyday of CND the call was clear and unambiguous: Britain had a hydrogen bomb, had demonstrated her ability to build and explode it; Britain still had international standing as a leader among nations; Britain ought to make the dramatic moral gesture of relinquishing the use of nuclear weapons, unilaterally if necessary. The debate that raged was intense and polemic on all sides (see Norman Moss, in Bibliography, p. 289, for an excellent summary of its stages). At the Labour Party Conference in 1960 the influence of CND won a vote endorsing unilateral nuclear disarmament. But nothing came of it. The movement began to diffuse, its focus spreading into increasingly generalized protest. The Aldermaston March became institutionalized, and gradually dwindled both in numbers and in interest. In the 1970s CND survives, but at a low ebb; the threat that spurred it into existence has also, apparently, been institutionalized, no longer recognized as a credible target for social dissent.

CND was from its outset a political, and soon thereafter an explicitly party-political movement. Its challenge was on the level of policy, of the abstract concepts of right and wrong – an organization whose commitment was first and foremost ethical. It did not much concern itself with the scientific or technical circumstances underlying the policy it was debating; it seems improbable that scientific or technical considerations could have exerted more than a moderate influence. It was not necessary to know how hydrogen bombs might wipe out humanity; to know that they could was sufficient basis for action.

But in St Louis, in the USA, a group of scientists were approaching the nuclear situation from a very different angle. To these scientists it was clear that the general public was baffled and frightened by manifestations like fallout and strontium-90, and that official responses were inadequate if not openly dishonest. In April 1958 the scientists, based at Washington University in St Louis, banded together as a Committee for Nuclear Information (CNI), and offered their services to the public, to answer questions, to give talks, and to tackle unresolved scientific problems which official science seemed to be dodging. CNI, however, explicitly declined to comment on the correctness or otherwise of official nuclear policy as such. Its brief was solely to clarify and explain the specifically scientific and technical aspects – medical, physical, chemical, biological – so members of the public could formulate their own opinions with a clearer understanding of the subject. It did not, in practice, work out quite as objectively as this. By implication, CNI commentary filled in the missing side of a story, half of which had been put forward by official sources; and CNI's side was likely to suggest to the public a viewpoint other than that espoused by officialdom.

In 1957 St Louis had become one of six US cities whose milk was sampled by the Public Health Service to check for strontium-90. The garbled flow of raw data and conflicting interpretations left the citizenry alarmed and confused. When CNI offered straightforward information as to what was known and not known about fallout, the city welcomed the offer. As CNI's reputation grew, its members moved farther afield. CNI data on iodine-131 around the Nevada test site suggested that children might have received many hundreds of times the recommended maximum radiation exposure as a result of tests in 1953 and later; when CNI presented this information to the Congressional Joint Committee the AEC rejected it completely, even though a report just prepared by one of the AEC's own staff, Dr Harold Knapp, agreed with the CNI testimony. We shall have more to say about the Knapp report shortly. In 1958 an American biochemist, Herman

T–F

Kalckar, published in *Nature* a paper pointing out that body levels of strontium-90 in children could be measured simply by analysing their baby teeth after they had been shed. CNI took on the task of organizing collection of baby teeth. From December 1958 the CNI Baby Tooth Survey got underway; by 1966 with the cooperation of St Louis mothers and children the Survey had collected over 200 000 teeth, analysis of which made possible the first detailed study of the absorption of strontium-90 from fallout by a large population of children.

CNI founded a monthly news bulletin, *Nuclear Information*, which in the early 1960s became *Scientist and Citizen* and in 1969 *Environment*, by which time its circulation was both national and international, as was its reputation for penetrating investigative science-writing. CNI, too, grew. Scientists based at other US universities banded together into similar information committees, and established a national link-up for exchange of data and coordination of policy. In the late 1960s the nationwide movement led to the founding of the Scientists' Institute for Public Information (SIPI), based in New York, with many affiliated organizations including the original St Louis CNI. In January 1973 SIPI, whose Advisory Board by now included an impressive array of leading scientists, among them Margaret Mead, Lamont Cole, and René Dubos, became publishers of *Environment*. Barry Commoner, one of the original St Louis CNI team, and later one of the leaders of the US environmental movement, became SIPI chairman. We shall have more to say later about the role of SIPI in nuclear controversy.

In November 1958 the USA, USSR and UK began a moratorium on nuclear testing which lasted nearly three years. The three years were not, however, all sweetness and light. France exploded its first nuclear weapon in February 1960. A planned summit meeting between President Eisenhower and Premier Khrushchev was aborted when a Soviet ground-to-air missile brought down a high-flying American U-2 spy plane over Soviet territory. In August 1961 came the Berlin crisis, and the building of the Berlin Wall. On 17 October 1961 the USSR announced that it was about to test a 50-megaton prototype of a 100-megaton weapon. The test duly took place on 30 October, with a yield estimated by the USA at 58 megatons; the open season for fallout was off with the loudest bang ever. One of the US tests reached new extremes, both of height and of lunacy. Despite outspoken worry by some knowledgeable scientists, which was severely hampered by US secrecy, the USA on 9 July 1962 detonated Project Starfish, a high-altitude nuclear explosion over the Pacific. The result of the explosion, at an orbital height, was semi-permanent dislocation of some of the earth's natural radiation belts, then only recently discovered, whose role in the atmospheric balance of the planet was far from well understood.

In October 1962 US intelligence discovered that the USSR was in the process of installing ballistic missiles in Cuba. The result was a showdown between US President Kennedy and Soviet Premier Khrushchev, during which a sizeable number of the people on earth feared that at any minute the final nuclear holocaust would descend.

The near miss prompted international cooperation; after the signing of the Partial Test Ban treaty on 31 July 1963 the tests went underground. But before they did the US testers managed to fit in one spectacularly messy one in Nevada. Called 'Project Rollercoaster', it was one of those intended to provide data on what would happen if a nuclear weapon should be involved in a non-nuclear accident, which detonated the high-explosive trigger without causing a fission explosion. The first 'Rollercoaster' shot on 15 May 1963 was aptly named. The high explosive filled the air with a cloud of plutonium, which rode the wind all the way from the Nevada site into California.

In any event, although the US and Soviet tests went underground after July 1963, it should not be inferred that their numbers decreased – on the contrary (see Figure 11). Then on 16 October 1964 the People's Republic of China set off their first nuclear explosion.

By 1962, as testing rose in a relentless crescendo, doubts were likewise arising even within the AEC. Dr Harold Knapp of the AEC's Division of Biology and Medicine had completed his lengthy and thorough study of the radiological history of iodine-131, from its creation in weapons tests via its deposition on grass, ingestion by cattle, reappearance in milk and eventual arrival in the thyroids of milk-drinkers, especially children. Knapp's report made mincemeat of earlier AEC assertions that exposure levels had remained insignificant. As might have been expected, the report did not delight the AEC. Since Dr Knapp was himself an AEC staff member the report was withheld while it was scrutinized by a specially-convened AEC committee of 'qualified scientists with specialized backgrounds' which became known as the Langham Committee.

Four of its five members were indeed qualified scientists; one was Dr John Gofman of the AEC's Lawrence Livermore Laboratory; all four were generally in favour of the Knapp report. The one dissenter was not only a non-scientist but also the chief of the Off-Site Radiological Safety Organization of the Nevada test site. Not surprisingly – since his was the responsibility for forestalling any untoward consequences of radioiodine from weapons tests – he was 'completely negative' about the nature and utility of the Knapp report. Because of his view the AEC could declare that the Committee's assessment of the Knapp report was 'not unanimous'. The report was finally published in 1963, in a third version, the first having been suppressed by the AEC and the second classified secret. The third version was published with the committee's 'not unanimous' review of it; on the day of publication Knapp resigned from the AEC. Dr Gofman's experience as a member of the Langham Committee was a key factor in converting him from a valuable senior staff man with the AEC into one of the AEC's most outspoken and relentless adversaries.

From the early 1960s onwards Dr Ernest Sternglass, Professor of Radiation Physics at the University of Pittsburgh, grew progressively more worried about the fallout from nuclear weapons testing, as manifest in statistics for infant mortality following tests. He was dealing with effects which were open to a broad range of interpretation, because of the obliqueness of the statistical inferences involved, and the smallness of the available statistical samples. What he found in the statistics indicated to him that even the low-level radiation associated with a rain-out of radioactivity like that over Troy in 1953 produced an identifiable increase in leukaemia in the children exposed. Official AEC response to his findings was not merely to deny their validity but to attempt to discredit Dr Sternglass, to cast doubts on his professional standing, and to endeavour to stifle any open discussion of his thesis. A biophysicist at the Livermore lab was requested to prepare a critique of Sternglass's assumptions. At an AEC-sponsored symposium at Hanford in May 1969 Dr Sternglass delivered a paper declaring that some 400 000 infants less than a year old had probably died as a result of nuclear fallout between 1950 and 1965. Also present at this symposium was the British radiobiologist Dr Alice Stewart; her classic investigation of childhood cancers, published in June 1958, had first demonstrated that the dose from diagnostic X-rays given to pregnant mothers could produce a detectable increase in the frequency of cancer in children so exposed. Such a dose, of a few rads or less, is comparable to the dose resulting from nuclear fallout of the kind deposited at Troy, St George and elsewhere.

As they had done from the outset AEC scientists at the Hanford symposium decried Dr Sternglass's work as scientifically indefensible. But the *Bulletin of the Atomic Scientists*, which had published in April 1969 an article by Dr Sternglass entitled 'Infant Mortality and Nuclear Tests', published in June another entitled 'Can the Infants Survive?'. *Esquire*, alerted by news stories, ran a special feature in September 1969, 'The Death of All Children', written by Dr Sternglass at the request of the editor, Harold Hayes. *Esquire* decided to publicize the feature with full-page ads in the *New York Times* and the *Washington Post*, and to send advance copies to every congressman and senator. Congress was at that time debating the advisability of embarking on the installation of an Anti-Ballistic Missile (ABM) system which would intercept any incoming nuclear missiles and wipe them out with US nuclear missiles. Clearly, if the Sternglass evidence was credible, the consequent blanket of fallout would probably have long-term effects as globally lethal as all-out nuclear war. However, within a fortnight Congress gave the go-ahead to the anti-ballistic missile system.

In October 1969 Dr Sternglass was invited to debate his thesis at Berkeley with the scientist who had prepared the AEC critique of his analysis, and whose article on the controversy was about to appear in the *Bulletin of the Atomic Scientists*. The AEC scientist's name was Dr Arthur Tamplin. Dr Tamplin had been engaged for a number of years in work at the Livermore lab, developing predictions of the biological pathways and ultimate destinations of radioisotopes from nuclear explosions. His credentials as a critic of the Sternglass analysis were clearly in order. However, the outcome of the AEC's assignment to Dr Tamplin was one of the most ironic misfires in the whole history of AEC public relations.

As the AEC had doubtless expected, Dr Tamplin did indeed disagree with Dr Sternglass. Dr Sternglass had postulated that fallout had caused the deaths of 400 000 children; Dr Tamplin went over the data and came to the conclusion that Dr Sternglass was overstating the case by a factor of at least 100. Unfortunately, from the AEC's point of view, this meant that Dr Tamplin himself considered that nuclear fallout had killed as many as 4000 children. Dr Tamplin's AEC superiors invited him to delete this unwelcome number from his critique; Dr Tamplin, backed by the Livermore lab's Associate Director, Dr John Gofman, refused to do so. At the Berkeley debate Dr Gofman – a senior AEC authority on the relation between radiation, chromosome defects and cancer – disavowed before the audience the official AEC position; he declared that to the best of his knowledge no data from animal experiments had any bearing on the low-level effects – small reductions in birth weight and in ability to fight infection – which the Sternglass hypothesis identified as the radiogenic reasons for increased infant mortality. As AEC pressure on Drs Gofman and Tamplin grew more overt, they found themselves, although still senior AEC staff members, in the forefront of the burgeoning challenge to AEC standards and procedures.

Into 1969 Dr Sternglass had been concerned almost exclusively with the effects of low-level radiation from nuclear weapons fallout; the Tamplin critique had also been directed to this end. But by this time another source of low-level radiation had begun to attract attention: the running releases of radioactivity from nuclear power stations and fuel-reprocessing plants. By the end of 1969 Drs Sternglass, Gofman and Tamplin found themselves increasingly preoccupied with the effects of civil nuclear systems on public health; and they were not alone.

In Minnesota, Northern States Power were building a nuclear station at Monticello, on the uppermost reaches of the Mississippi River, a 545-MWe boiling water reactor. The Minnesota Pollution Control Agency, a state authority, requested from the utility some details about its proposed discharges into the Mississippi, and received a curt refusal. The utility did not take the trouble to explain that its radioactive discharges were the responsibility of the AEC, at the Federal level, and outside the jurisdiction of the State. In 1967 the MPCA asked the help of Dr Dean Abrahamson of the University of Minnesota. Like Dr Gofman, Dr Abrahamson was both a qualified doctor of medicine and a qualified nuclear physicist; Dr Abrahamson and his colleagues – who in due course formed the Minnesota Committee for Environmental Information – were soon embroiled in a head-on collision between the AEC and the State of Minnesota over radiation standards. State authorities engaged Dr Ernest Tsivoglou of the Georgia Institute of Technology to report on standards. In March 1969 Dr Tsivoglou recommended that the state set discharge standards in line with performance figures given by the utility and the builders, General Electric, for discharge of radioactive wastes – standards which would thereby be some fifty times lower than the national standards set by the AEC. The AEC declared that Minnesota had no right to set its own standards, safer or otherwise; and the utility took the State to court.

With AEC backing, the utility in due course won its case, but the episode did not go unnoticed. Local communities about to become hosts to large new nuclear stations evidenced mounting concern about radiation hazards. Their concern was furthermore no longer theoretical, but was manifesting itself in legal challenges to licensing of the proposed facilities. At the end of 1969 the Congressional Joint Committee held a series of hearings into the environmental effects of producing electrical power, eliciting a massive body of internally contradictory testimony from interested spokesmen on every aspect of the question, especially the nuclear. Areas in particular doubt included operating safety (see p. 193ff), and radiation hazards from normal operation. The vague uneasiness of the lay public was now reinforced by expert opinion of undeniable authority, particularly that of Drs Gofman and Tamplin. It was no longer possible for the nuclear establishment to shrug off objection by claiming that objectors were unqualified.

On 3 December 1970 the AEC published amendments to its guidelines on radiation standards. Following hearings in January and February 1971 it put forward proposed revisions of basic standards for the control of radioactive releases from nuclear power plants, to keep these releases 'as low as practicable'. The effect of the revision was in essence to do what Drs Gofman and Tamplin had suggested: to lower by a factor of about one hundred the amounts of radioactivity which were permitted to be released at the site boundary of a nuclear reactor plant. Certain radioisotopes, notably radioiodine, were brought under yet more stringent control, their permitted releases being reduced by a factor of 100 000. To be sure, the 'as low as practicable' criteria simply acknowledged that routine releases could, with current technology, be kept this low without difficulty. But the AEC was, as Dr Gofman commented, 'responding sensibly to sensible pressures'. The pressures were not, however, about to relax.

In any case, it was clear that civil nuclear systems were not in the same league as nuclear weapons when it came to environmental radioactivity. On 18 December 1970, a 20-kiloton nuclear explosion called 'Baneberry' scattered radioactivity over twelve western states and across the US border into Canada – which contravened the Partial Test Ban Treaty. The 'Baneberry' shot was detonated 300 metres underground at the Nevada Test Site, but it re-emerged, venting a cloud of fallout nearly three kilometres high, forcing the evacuation of 600 workers from the site and contaminating 300 of them, as well as 80 cars. A work camp close to the site of the shot was so badly contaminated it was unsafe for occupancy for two months. After 'Baneberry' the AEC suspended nuclear testing to examine safety standards. But by mid 1971 they were at it again.

Nonetheless, the AEC was by now getting an increasing volume of complaints by Nevadans about the operations at the Nevada site. So the AEC looked for a test site that would not upset folks at home, but would likewise not upset foreigners. The choice they made upset both, more than any previous test episode. For a 5-megaton underground test called 'Cannikin' the AEC settled on the Aleutian island of Amchitka – in one of the most active seismic regions of the entire planet. By the time the test was fired the shot was heard round the world. Protests came from almost every circum-Pacific country, including Canada, Peru, New Zealand, Australia, and Japan. A boatload of determined opponents in a tiny trawler rechristened the *Greenpeace* tried to sail into the test zone. The Canadian government, spurred by Canadian objectors numbering into the millions, demanded that the test be cancelled. Within the USA objectors carried their protest all the way to the Supreme Court which – by a majority of four to three – refused to intervene. The Senate Majority leader called the test 'dangerous and an outrage', and thirty-five Senators asked President Nixon to stop it. But on 6 November 1971 'Cannikin' went off. Experts declared that the earthquake that hit Hokkaido and Honshu in Japan the following day had nothing to do with the test. Alaska authorities estimated that more than 300 sea-otters had been killed. The radioactivity released by the blast did not put in an immediate appearance in the adjacent air or water; but objectors noted that an area which sustains dozens of earth tremors annually was hardly a satisfactory resting-place for many millions of curies of long-lived radioisotopes.

The feelings which had launched the first *Greenpeace* were meanwhile being redirected southward. Since 1966 France had been carrying out nuclear tests in the atmosphere above Mururoa Atoll, about 1600 kilometres south-east of Tahiti. In August 1971, asserting that that radioactivity from the French tests was now detectable in the snows of the Andes, Peru warned France that diplomatic relations would be broken off if France persisted in the tests. As the 1972 test series was about to begin another Canadian boat, the ketch *Greenpeace III*, was rammed by a French naval vessel in international waters just outside the test zone. The United Nations Conference on the Human Environment passed by fifty-two votes to none a resolution put forward by New Zealand and supported by almost all the circum-Pacific countries – except the USA and China – condemning nuclear weapons tests. Australian and New Zealand trades unions boycotted French shipping and airlines. But on 28 June the first test took place nonetheless. The same pattern followed in 1973, except that this time the French navy boarded the *Greenpeace III* and beat up the crew, severely injuring the Canadian skipper. Despite international outcry France carried on in 1974 as usual, although announcing plans to go underground for later tests.

To anyone who might have felt that, in a nuclear context, a bomb is a bomb, another activity of the bomb-makers presented further confusion. In an attempt to find gainful non-military employment for nuclear bombs, the AEC in July 1957 set up the Division of Peaceful Nuclear Explosives. Its programme was given the Biblical appellation 'Plowshare', as something into which nuclear swords could be beaten. The prospect of carrying out civil engineering blasting with nuclear devices looked to be economically promising, especially for very large-scale projects for which ordinary chemical high explosives would be prohibitively expensive. There were, of course difficulties. For one thing, the point at which a single nuclear explosive device becomes less expensive than the equivalent chemical explosive is upwards of 1000 tonnes of TNT. The opportunities for single blasts of this magnitude for civil engineering is somewhat limited. More awkward still is the fact that a nuclear explosion, however peaceful, produces a belch of fission products which will then hang about the place making it slightly unhealthy.

Early Plowshare proposals ignored the problem of residual radioactivity. One of the first plans to be seriously mooted was Project Chariot, a scheme to blast out a harbour on the northern coast of Alaska by nuclear explosive. But investigations by scientists associated with the Committee for Nuclear Information revealed that the resultant fallout would accumulate on Arctic lichen, be eaten by reindeer and caribou, and be concentrated enough to present a significant radiological hazard to Eskimos who then ate the animals. When CNI scientists in Alaska explained their findings to the local people the response persuaded the AEC to abandon the project. A similar makework programme was the grandiloquent plan for a new canal across central America, somewhere south of the Panama Canal, which was said to be too small for large tankers and which was in any case in an area the USA was finding ever harder to handle politically. The new nuclear canal would require the simultaneous detonation of about 250 megatons of nuclear explosives – a Plowshare to reckon with, in any terms. Evicting thousands of native people from the surrounding area was accepted as an inconvenience, as was the likelihood of lingering radioactivity which would make it unwise for them to return for years afterwards. But after six years of work between 1964 and 1970 the Atlantic–Pacific Interoceanic Canal Study Commission reported that the best solution would be to use ordinary, manageable amounts of conventional explosives to widen the existing Panama Canal.

By 1969, despite the enthusiasm of advocates like Dr Edward Teller, the possibilities for Plowshare applications were being narrowed down to those which would – at least theoretically –

confine the resultant fission products rather than releasing them to the atmosphere. (The USSR had fewer scruples, and after earlier Plowshare-type explosions in the late 1960s is still proposing in 1975 to use peaceful nuclear explosions for the diversion of rivers and other applications of planetary engineering. The side-effects – not only radiological – continue to worry many observers.) The most persuasive proposals for US applications were those calling for underground explosions, to create huge chambers for storage, especially of dangerous wastes, and to fracture gas-bearing rock-formations. No industrial interests have thus far proved willing to support underground storage excavation by nuclear explosives; but so-called 'gas-stimulation' has become a bitterly controversial activity, especially in Colorado.

The first underground gas-stimulation explosion in the Plowshare programme took place in January 1967 near Farmington, New Mexico; it was called 'Gasbuggy'. The next was called 'Rulison', a 43-kiloton explosion north-west of Grand Junction, Colorado, in September 1969. Neither of these 'experimental' explosions produced gas of very high quality. What gas was produced has mainly been flared at the top of gas wells drilled into the shattered formations. Local objectors to the explosions have also been unhappy about the subsequent flaring, pointing out that the gas contains radioactive tritium, which is thereby being disseminated into the Colorado atmosphere. They are even more unhappy at the thought that radioactive gas may be fed into the gas-mains, to emerge faintly radioactive from household cookers.

The third project, called 'Rio Blanco', drew the most outspoken opposition yet. It was a scheme utilizing three nuclear bombs, one above the other in a single shaft, intended to create a huge broken cavity through which the gas could readily permeate to wells. 'Rio Blanco' was fought through the courts; but in due course, on 17 May 1973, the blast was triggered. It did not, apparently, accomplish its end; the three separate cavities did not join up. Over $40 000 in claims have already been paid out for damage occasioned by the shock-wave

('Rulison' required a total payout of $155 676); structural bracing in the vicinity of the 'Rio Blanco' shot cost another $142 500. The nuclear gas-stimulation programme would, says the US General Accounting Office, involve drilling some 5680 wells and setting off 29 680 nuclear explosions in the coming thirty-five to sixty-five years.

Once created, radioactivity is an environmental liability which may endure an unthinkably long time. Creating it under uncontrolled conditions, especially in the atmosphere, is an act of unconscionable biological brutality. Even when it is created in confined conditions it has an unhappy knack of escaping. If we are to continue to create radioactivity in ever-increasing quantities, we are going to have to improve our vigilance manyfold, or risk irreparable damage to the fabric of life on earth.

6. Reactors Off and Running

Decision-making in nuclear matters has from the outset been the prerogative of government. So long as nuclear activities were directed entirely towards military ends this was clearly appropriate. However, as civil applications of nuclear energy came more and more to the fore, the relationship between government, the scientific community, industry and commerce became progressively more complex. The precise details differed from nation to nation, as we shall see. But the collaborative arrangement developed in a form without effective parallel in any other field of human endeavour, especially as the nuclear fuel cycle, sector by sector, began to exhibit a civil as well as a military side.

The first nation to emphasize the civil rather than the military aspects of nuclear activities was Canada. As the third partner, with Britain and the USA, in the Manhattan bomb-development project, Canada was chosen as the site for construction of major experimental facilities, including the first reactor outside the USA. But after the Second World War Britain recalled most of her scientists from Canada to undertake the British nuclear programme. Thus bereft, the Canadian government could see no point in a Canadian programme of nuclear weapons; this attitude has persisted ever since. Nonetheless Canada found itself with the beginnings of a nuclear establishment at Chalk River, some 200 kilometres north of Ottawa, whose construction was well under way when the British left.

Canada's first two reactors, built at Chalk River, were the small ZEEP reactor and the considerably larger NRX, which went critical in 1947 and reached its full power of 40 MWt in May 1948. The NRX was a research reactor, in some respects the first precursor of the CANDU. In common with most heavy water designs the NRX was an efficient producer of plutonium, which Canada sold to the USA and Britain, a procedure which has been followed and expanded with later Canadian reactors. But the work at Chalk River focused mainly on using the NRX for a wide range of basic research. It soon established itself as the most successful experimental reactor in the world. Then, ironically, the one country which after the Second World War had decided to develop its infant nuclear capacity purely for research was the first to play host to a major reactor accident.

On 12 December 1952 a technician in the basement of the NRX building erroneously opened three or four valves (the exact number has never been established) which lifted three or four of the reactor's twelve shut-off rods out of the core. The supervisor, seeing the red lights come on at the control desk, left his assistant in charge and went to see what was going on. In the basement he at once realized what had happened, reset the valves and telephoned his assistant to press buttons 4 and 3 to restore normal operation. In his haste he inadvertently said '4 and 1'; before he could correct himself his assistant had laid down the phone and complied. Unknown to either the supervisor or his assistant, resetting the valves had extinguished the red lights without in fact fully reinserting the rods; the assistant had no reason to question the safety of pressing button 1. Doing so lifted four more shut-off rods out of the core; the power level of the reactor began to rise. Within twenty seconds the assistant realized that all was not well, and pushed the scram button. This should have reinserted all the shut-off rods; but it did not. Only one of the seven or eight withdrawn rods dropped in, and it did so very slowly, taking some ninety seconds to fall just over three metres. The operators decided that it would be necessary to dump the heavy water out of the calandria, a last-ditch emergency provision to shut off the fission reaction. The heavy water took thirty seconds to drain out of the tank, and instruments showed that the reactor power level had fallen to zero.

But in the basement the supervisor and another staff member could see through a doorway water pouring out of the system. They rushed in with a bucket, thinking it was heavy water – but it was in fact light water coolant, and radioactive. Up above there was a rumble and water jetted out of the reactor. Radioactivity alarms began to sound, both in the reactor building and in the chemical extraction building on another part of the Chalk River site. Sirens warned the site personnel to take refuge indoors; a few minutes later came the top-level order to evacuate the entire plant. Only the control room staff remained behind, donning gas-masks. From beginning to end the whole accident sequence had lasted only seventy seconds.

The unintentional lifting of the control rods had let the chain reaction speed up to such an extent that the heat release had melted some of the uranium fuel. The fission energy itself did not produce an explosion. But the heat boiled some of the coolant, forming steam bubbles which were much less absorptive of neutrons, allowing the fission reaction to speed up still more. Within a few seconds the melting uranium fuel and aluminium cladding began to react with water and steam; the hot uranium metal stripped the oxygen from water molecules, leaving free hydrogen; the hydrogen mixed with inrushing air entering through ruptured piping, and the resulting explosion heaved a four-tonne helium gasholder to the top of its travel and jammed it there.

The surge of temperature and pressure, the chemical reactions and the explosions pretty much demolished the reactor core, and disgorged radioactivity in all directions. About 10 000 curies of long-lived fission products were carried into the basement by the leak of 4 million litres of cooling water. Fortunately the emergency procedures at the plant were effective. It was subsequently reported that no staff received excess radiation exposure during the accident itself, and that during the clean-up – a protracted and messy business, lasting many months – the highest dosage received was only 17 roentgens, with most others being below 4 roentgens. Although well above recommended levels these are in the circumstances comparatively modest exposures; when it is noted that the accident sequence included an almost complete failure of the scram-rod system, the reactor staff may even count themselves fortunate.

Needless to say the lingering radioactive contamination presented the most awkward obstacle to those charged with tidying up the shambles. They can in fact be credited with a coup of sorts: they managed in due course to devise a method whereby they could extract the entire contaminated calandria from the interior of the reactor shielding and replace it with a new structure. Few of those at Chalk River on the evening of 12 December 1952 would ever have expected to see the NRX back in service after its resounding hiccup. But in only 14 months the NRX was back – in plenty of time to take over again later in the 1950s while its 200-MWt successor, the NRU, underwent two months of extensive decontamination. The decontamination became necessary on 25 May 1958 when an irradiated fuel element broke and caught fire inside the NRU refuelling machine. At one point a 1-metre length of fiercely radio-active fuel fell out of the refuelling machine and burned. Fortunately it had landed in a maintenance pit; the radiation dose-rate in the pit was estimated as high as 10 000 roentgens per hour. Some 600 men were involved in the clean-up, and 400 000 square metres around the NRU building were contaminated.

Like the Canadians, the French participated in the early stages of the Manhattan project, and were then gradually eased out of the picture. Like the Canadians, the French immediately after the Second World War set up a government nuclear undertaking directed to pure research: the Commissariat d'Energie Atomique (CEA). Like the Canadians the French found themselves with useful reserves of uranium to be mined. However, unlike the Canadians, the French began to wonder by the mid 1950s whether it might not be advisable to proceed with development of nuclear weapons.

Their first research reactor, Zoë or – less poetically – EL-1, went critical in 1948, at the research centre at Fontenay-aux-Roses near Paris. By 1952 the first French nuclear power programme had been drawn up. The air-cooled, graphite-moderated G-1 reactor at Marcoule, a 40-MWe dual-purpose plutonium–power reactor went critical in 1956, like its British cousin at Calder Hall. As in Britain, the first commercial nuclear stations in France were offspring of the plutonium-production reactors. The 70-MWe Chinon-1 station built on the Loire for Electricité de France went critical in September 1962.

After the start-up of the Obninsk 'first atomic power station' APS-1, and with the gradual relaxation of East–West tension, the Soviet nuclear effort also expanded from military into civil applications. The first full-scale Soviet nuclear power station was constructed at Troitsk in south-western Siberia, and came on stream in 1958. Its reactors, eventually six in number, were 100-MWe descendents of the Obninsk plant, a distinctive Soviet design using a graphite moderator enclosing pressure tubes filled with light water coolant. Work also began on a Soviet design of pressurized water reactor. The first 265-MWe PWR started up at Novovoronezh in October 1963.

The United Kingdom Atomic Energy Authority was created on 1 January 1954, and a government White Paper published in February 1955 laid down the basis for a programme of civil nuclear power in Britain. The White Paper pointed out the anticipated growth in demand for electricity in Britain, the inability of the coal industry to keep up with this demand, and the long-term likelihood that nuclear electricity would prove cheaper than coal; it made no attempt to argue that nuclear electricity would be instantly competitive with coal, but asserted that Britain's lead in technology could not be allowed to dwindle. On the basis that the costs would be comparable to those of electricity from coal the government gave the go-ahead for a programme of twelve nuclear stations to be constructed throughout the ensuing decade. This programme was later twice revised, in October 1957 and then in June 1960, into a less ambitious one. But even before Calder Hall had started up, the Central Electricity Generating Board had begun to order the first generation of commercial Magnox stations, beginning with the twin-reactor stations at Berkeley in Gloucestershire and Bradwell in Essex.

On 8 October 1957 the physicist in charge of the Windscale Number One plutonium production reactor threw a switch too soon. He was carrying out a routine operation known as releasing

Wigner energy, which involved raising and lowering the power level. According to his instruments he deduced that the core temperature was falling, without completing the desired Wigner release. He did not have a Pile Operating Manual, with its special sections on Wigner release, to help him, nor had he sufficiently detailed instructions. The physicist nonetheless decided to give the power level another short boost, to bring the temperature back up and complete the Wigner release. What he did not know was that thermocouples recording core temperatures were not in the hottest part of the core. Core temperatures at some points were considerably higher than the physicist realized. When at 11.05 a.m. he withdrew control rods to raise the power level again, the resulting additional temperature rise eventually ignited at least one fuel rod.

The physicist had no idea that anything was amiss. Not until 5.40 a.m. on 10 October – 42 hours 35 minutes later – was there any external sign that all was not well inside the core of the Windscale Number One reactor. Then instruments began to show that radioactivity was reaching the filters on the top of the cooling-air discharge stack. The filters were known as 'Cockcroft's Folly'; Sir John Cockcroft had insisted that they be installed, after the stack had been built, as a precautionary measure – to the derision of some of his colleagues. As it turned out, 'Cockcroft's Folly' probably kept a major accident from becoming a catastrophe. By the time the Windscale staff realized that something was wrong, the fire was an inferno, and spreading fast.

Unfortunately it was far from clear what could be done about it. Molten uranium and cladding, steeped in fission products, burned fiercely in about 150 fuel channels, fanned by the onrush of air which had by this time no hope of cooling the core. The graphite, too, was aflame. Tom Tuohy, later the Windscale General Manager, recalled standing on the pile cap wearing breathing apparatus, looking down through a viewing port above the cooling pond, and seeing flames shooting out of the discharge face of the core and playing against the concrete shielding of the outer wall – concrete whose specifications required that it be kept below a certain temperature, lest it weaken and collapse. At the height of the fire eleven tonnes of uranium were ablaze.

The Windscale staff knew only too well that water and molten metal in contact might react, oxidizing the metal and leaving hydrogen to mix with incoming air and explode. No one could be sure that such an explosion would not rend open the shielding, disgorging a hell-cloud of scorching radioactivity. Some of the staff insisted that they must first try carbon dioxide, despite Tuohy's remorseless reminder that – at the fire's temperature – the oxygen of carbon dioxide would feed the flames as effectively as air.

An ICI tanker-load of fresh liquid carbon dioxide coolant for the Calder Hall reactors had just arrived on site. But Tuohy's prediction proved all too correct; fed with carbon dioxide the flames only intensified. Water was the only recourse. In the early hours of Friday 11 October the decision was made: the Chief Constable of Cumberland was warned of the possibility of an emergency. Firehoses were hauled up the charge face of the reactor. Their nozzles were cut off, and the hoses were instead coupled to the entry ports on a line of fuel channels about a metre above the heart of the fire. By this time the fire had been raging out of control for more than twenty-four hours. Tuohy ordered everyone else out of the plant but himself, one colleague, and the local fire chief. At 8.55 a.m. they turned on the firehoses.

It worked. Slowly the fire subsided and died. But the Windscale troubles were far from over. The staff had wrestled with the fire for more than a day before word reached the press and public, including local people, that something was wrong at Windscale. Even while Tuohy and his staff were trying to conquer the flames it was clear that the fire had released a vast cloud of radioisotopes from the melted fuel. The stack filters had trapped a large proportion of the escaping radioactivity, but by no means all. Outside the plant the question was – how much radioactivity had belched out of the stack and descended over Westmorland and Cumberland? Had it gone farther? What kind was it, and how dangerous? Most important of all – what had to be done, and done quickly?

One radioisotope above all was rapidly identified as the most hazardous – iodine-131, with its short half-life, high activity and instinct to home in on the human thyroid. (In due course it was estimated that some 20 000 curies of iodine-131 had been released *to the atmosphere.) Decisions were made. Cattle grazing in fields where the radioisotopes had descended would produce milk laced with radioiodine; such milk must not be drunk. By arrangements between the Atomic Energy Authority, local police, the Milk Marketing Board, and the Ministry of Agriculture, Fisheries and Food, milk from an area of more than 500 square kilometres – some 2 million litres – was instead poured into rivers and the sea. It was said locally that the worst after-effect of the Windscale fire was the sour stench of every waterway for weeks afterward. Farmers were compensated by the government; it was also said locally that, to judge by the amount of claims, local cattle must have been yielding more milk than any other cattle in the country.*

It has never been explained why the government opted for the dramatic gesture of pouring away milk, instead of merely drying it and storing it for a few weeks until the radioiodine had decayed away. It must be assumed that the government anticipated a public outcry at any attempt to let 'radioactive milk' return to the market. It is also reasonable to suppose that the government, with the grandiose gesture of pouring away the milk, thought to distract attention from the other possible consequences which might only become manifest many years later. There does not appear to have been any effort to keep track of people who were near Windscale during the fire. It took place nearly two decades ago; in the mid 1970s there remains only the lingering local suspicion that a lot of people seem to die of diseases like cancer these days. There may well be absolutely no medical or statistical bases for such rumours. The evidence one way or the other seems not to have been considered worth collecting.

In the aftermath of the Windscale fire the Windscale Number Two production reactor was shut down while inquiries were carried out. The full report of the inquiry was never published; the published version made it clear that design changes to the Number Two reactor to prevent a recurrence of the fire would be prohibitively expensive. Both reactors were in due course plugged with concrete and entombed. Fortunately the Magnox stations – the Calder Hall and Chapelcross military installations, and the commercial plants of the new civil programme – operated at a tem-

*perature high enough to obviate the need for Wigner release. The
Windscale fire was a once-in-a-lifetime event. For those involved,
many of whom have become senior figures in the British nuclear
establishment, once was enough.*

In 1959 Parliament passed the Nuclear Installations (Licensing and Insurance) Act, creating the Inspectorate of Nuclear
Installations, responsible for the safety of commercial nuclear
power stations and research reactors. From the moment a
nuclear power station was proposed, throughout its design and
construction, its operating lifetime and ultimately its decom-
missioning, the Nuclear Inspectorate was to be involved – in
effect as a technically qualified representative of the public.
The Act also introduced provisions severely limiting third-
party liability in the event of a nuclear mishap – scarcely a vote
of confidence in the effectuality of the newly-established
Inspectorate.

While the Atomic Energy Authority was building Calder
Hall and Chapelcross it was also building its remote Dounreay
installation in the northernmost tip of Scotland; the Dounreay
Fast Reactor went critical on 14 November 1959. The first
reactors at the CEGB's Berkeley and Bradwell Magnox
stations went critical in August 1961. The 32-MWe Windscale
AGR, the first of its kind, went critical in August 1962, as the
Authority continued to develop different reactors. In later
years the Berkeley and Bradwell stations were to prove
mainstays of the CEGB system.

In the USA under a joint project financed by the AEC, the
Duquesne Power & Light Company of Pennsylvania acquired
the Shippingport nuclear power station, the first nuclear power
station in the USA. The Shippingport station – using a trans-
planted naval reactor – was not expected to pay its way. As
indicated in Chapter 5 (p. 129), the USA had an abundance
of oil and gas at prices certain to undercut the nuclear costs, but
it was determined to stay abreast of nuclear developments in
Europe. The fanfare over Calder Hall put many American
nuclear noses out of joint.

In 1955 the AEC established a Cooperative Power Reactor Demonstration Program, offering substantial government finance to utilities prepared to join the AEC in building nuclear-powered generating stations. Despite the financial enticements, the utilities' lack of enthusiasm was in the event fully justified. Of the early experimental AEC-backed stations only the mainstream light-water designs – Shippingport, Dresden 1, Yankee Rowe, Indian Point 1, Big Rock Point, Humboldt Bay, La Crosse – and the little Peach Bottom HTGR survived. All the others were soon shut down, because they did not work, because they cost too much, or because they proved to be untrustworthy neighbours. Enrico Fermi 1 (sodium-cooled fast breeder), Hallam (sodium–graphite), CVTR (pressurized heavy water), Piqua (organic moderator) and three boiling water reactors (Elk River, Pathfinder and BONUS) lasted from criticality to shutdown at most eight years, and in one case (Hallam) only two. In the USA, more than in any other country, the gestation period for nuclear power was protracted and uncomfortable. Curiously enough it was the EBR-1, the world's first source of nuclear electricity, which set the stage for the nuclear struggles in the USA.

The EBR-1 had been given, so to speak, a core transplant; but the new core was misbehaving. It seemed to have a 'positive temperature coefficient of reactivity': a rise in core temperature provoked an increase in reactivity – which in turn reinforced the rise in temperature. For obvious reasons, a positive temperature coefficient – like any positive reactivity coefficient – presents problems of control for a reactor design, if not indeed safety problems. The National Reactor Testing Station staff decided to carry out a test on the EBR-1 Mark II core, which would involve interrupting the flow of sodium–potassium coolant, allowing the temperature to rise. The test was duly carried out, on 29 November 1955. Unfortunately, when the reactor operator went to shut down the reactor he mistakenly used the slow-acting control rods instead of the scram rods. The core temperature soared, to over 1100°C. The fuel, highly enriched uranium-235 in stainless steel cladding, softened and melted, slumping to the bottom of the containment, where inflowing coolant solidified it again, forming a cup which caught more of the melting fuel. (It was later found that 40 to 50 per cent of the fuel had melted.) The slow rods shut down the reaction, but the core was destroyed. No one on the scene received any radiation exposure. Neither, it then developed, did the accident receive any exposure. Not even Lewis Strauss, Chairman of the AEC, heard about it, although he might have been expected to be interested.

If Strauss had not been interested for any other reason he might have wished to meditate on a meeting a year before, on 10 November 1954 – proceedings that were still classified secret by the AEC at the time of the EBR-1 meltdown. Among the top US physicists present at the 1954 meeting, in the offices of Detroit Edison, were Walter Zinn, designer of the EBR-1, and Hans Bethe, a physicist of international repute. They were there to consider some unresolved questions about the performance of fast reactors. One of the most pressing questions related to possible meltdown in a reactor core composed of undiluted fissile material, like the core of the EBR-1.

In a thermal reactor, the fissile material is diluted by so much that is not fissile – uranium-238, structural material, moderator, coolant – that the operating configuration of the reactor fuel is just about optimum for reactivity. Any distortion of the core makes it less rather than more reactive. The same could not be said with confidence of a core devoid of moderator and made of highly concentrated fissile materials, pure uranium-235 or plutonium. The physicists at the Detroit Edison meeting had to concede that, so far as they knew, a fast reactor core which melted down might collapse into a yet more reactive configuration – possibly introducing so much additional reactivity that the full insertion of control rods would not suffice to shut off the chain reaction. The result might be a runaway, a 'disassembly' – that is, a small nuclear explosion.

The possibility of a runaway could not be discounted. To some of those present the only course open would be to build the first such reactor of any size in a remote location, to minimize the consequences should anything go amiss. The British did this, siting their first sizeable fast reactor on the north Scottish coast. But Detroit Edison, hosts to the US meeting, had good reason to oppose this view; for they were proposing that the AEC should help them build a fast reactor power station not far from Detroit. They would of course, they said, enclose the reactor in a containment strong enough to withstand any conceivable explosion and confine any release of radio-activity. In any case such an accident was highly improbable.

The EBR-1 meltdown only a year later disproved the last comfortable assumption; but some of the most obviously interested parties did not find this out for some months. It was not until 5 April 1956, that the *Wall Street Journal* put a direct question about the EBR-1 meltdown to Strauss as AEC Chairman, to which Strauss replied that it was 'news to him'. Not until that evening, at Strauss's personal behest, did the AEC issue a press release admitting that the meltdown had occurred. It cannot have been particularly welcome news to Detroit Edison, who were by this time leading a consortium of some thirty-five utilities and manufacturers, the Power Reactor Development Corporation, whose objective was to build Detroit Edison a prototype fast reactor power station. The Advisory Committee on Reactor Safeguards may have taken note of the EBR-1 episode; on 6 June 1956 they submitted a report declaring that not enough was known to guarantee public safety if such a plant were to be operated near an urban centre. Still, nothing loth, the Power Reactor Development Corporation duly filed with the AEC on 4 August 1956 an application for a construction license for a 300-MWe fast reactor station, to be sited at Lagoona Beach on Lake Erie, near Monroe, Michigan, about half-way between Detroit and Toledo – within forty kilometres of these cities and of Ann Arbor, and within fifty kilometres of Windsor, Ontario, across the Canadian border. The proposed station was to be named after the creator of the first reactor, Enrico Fermi.

The proposed Fermi plant was not greeted with delight by the inhabitants of the near-by cities. If they had known of the doubts expressed by the Advisory Committee on Reactor Safeguards they would have been even less enthusiastic. But the sceptical report was not made public; on the contrary, according to Congressman Chet Holifield of the Joint Committee on Atomic Energy, it was forthwith suppressed. In the light of their later pronouncements it is difficult to believe, twenty years on, that the opposition to construction of the Fermi plant was urged on by Holifield, by his colleague on the Joint Committee, Senator Clinton Anderson, and by AEC Commissioner James Ramey. But such was indeed the case. Anderson carried out a telephone campaign to get the United Auto Workers under Walter Reuther to seek an injunction to prevent construction of the Fermi plant. Reuther charged his lieutenant Leo Goodman to lead the opposition, and Goodman in turn persuaded Reuther to support the challenge to Detroit Edison with $350 000 of the Union's funds.

The first Geneva Conference on the Peaceful Uses of Atomic Energy in 1955 had already tackled the tricky subject of reactor safety. On 6 July 1956, at the behest of the Congressional Joint Committee, the AEC instructed a team of its experts, most of them based at the Brookhaven National Laboratory, to prepare a detailed analysis of the possibilities and public health implications of reactor accidents. The Committee was concerned lest uncertainty about liability in the event of accident deter utilities from building nuclear stations; we shall have more to say about this in a moment. The analysis, document number WASH-740, entitled *Theoretical Possibilities and Consequences of Major Accidents in Large Nuclear Power Plants*, was published in March 1957. The AEC apparently expected that the report, which stressed the extreme improbability of such accidents, would help to pacify the clamour around the western end of Lake Erie. If so, the AEC's expectation miscarried drastically.

For some reason lay readers were inclined to overlook WASH-740's estimates that the likelihood of a major accident was between 1 chance in 100 000 and 1 in 1000 million per year per reactor. Since the estimates were based on an almost total lack of actual operating experience, their failure to impress their readership may be understandable. What did catch the eye was the scale of the consequences anticipated should one of these apparently highly improbable occurrences nonetheless occur: in the worst case, 3400 deaths, 43 000 injuries, and property damage of $7000 million. This 'maximum credible accident', as it was called, occurred under the following assumptions: the reactor was of up to 200 M We output, its core nearing time for refuelling and accordingly containing its largest inventory of fission products; it was situated fifty kilometres from a city with a population of one million people; the accident envisioned a breach of the reactor containment sufficient to release one-half of the core inventory of fission products to the outside surroundings, at a time when the wind would blow the radioactive cloud in the direction of the population centre. But this combination of circumstances, leading to such appalling statistics, was not the worst that could conceivably arise.

The research and development arm of the Power Reactor Development Corporation commissioned the Engineering Research Institute at the University of Michigan to prepare an analysis with a brief broadly similar to that given the Brookhaven group. The Institute's report, focusing specifically on the Detroit Edison proposal, was published four months after WASH-740, in July 1957; its contents made even WASH-740 look like soothing bedtime reading. Unlike WASH-740 it assumed a complete rupture of the reactor containment, presumably accompanied by an explosion, such that the entire core inventory of the reactor was spewed into the air. The weather conditions were assumed to include a temperature inversion, preventing vertical mixing of the air, and a wind blowing slowly but steadily in the direction of Detroit. On this basis, the report estimated that 133 000 people would receive radiation doses of at least 450 roentgens – presumed fatal; 181 000 would receive 150–450 roentgens – causing immediate injury; and 245 000 would receive 25–150 roentgens – likely to cause long-term bodily or genetic injury. The report's authors, understandably, made no attempt after this to estimate property damage.

Like WASH-740 the Institute report emphasized that such an accident was a very remote probability indeed. This putative remoteness, however, juxtaposed with the astronomical numbers should it nonetheless come to pass, set the insurance business an unparalleled poser. Precisely this uncertainty had prompted the Congressional Joint Committee to direct the preparation of WASH-740. But they did not wait to see it; shortly after so directing, Committee members Melvin Price and Clinton Anderson put forward a bill to both Houses of Congress which has for two decades enshrined their names in nuclear annals as the Price–Anderson Act.

The Price–Anderson Act became law in 1957. Its purpose was uncomplicated, as were its provisions. In essence, it amounted to this: 'Private utilities will not build and operate nuclear stations if they may be bankrupted by claims arising from a major reactor accident. Therefore, let a utility be instructed to purchase from private insurers as much coverage as they will sell, against nuclear third-party liability. Thereafter the government will kick in an additional $500 million from Federal funds. Beyond this total there shall be no further financial liability.' That is, in the event that the $7000 million property damage foreseen as worst possible by WASH-740 should some black day occur, a maximum of – as it turned out – $500 million from government plus $60 million from private insurers would be available. Claims for the remaining $6440 million would not be entertained.

Notwithstanding WASH-740, the Price–Anderson Act – coupled with the veiled threat that the AEC would itself enter the electricity business – persuaded Commonwealth Edison, Consolidated Edison, Pacific Gas & Electric and other major utilities to take their first tentative steps along the nuclear road. In partnership with the AEC they commenced construction of the first generation of what were to become the world's best-selling reactors, the pressurized water reactor and boiling water reactor. The Dresden 1 BWR went critical in October 1959, the Yankee Rowe PWR in August 1960 and the Indian Point 1 PWR in August 1962.

By the end of the 1950s nuclear experience was no longer the exclusive property of the nuclear weapons nations and their Second World War partners. To be sure, such international traffic in nuclear matters had a rocky start. There was, once upon a time and briefly, a United Nations Atomic Energy Commission. It was set up by the UN General Assembly in January 1946, by the first UN resolution, regarding nuclear energy. It did not, however, last long, or accomplish much. Then on 8 December 1953, as the Cold War eased, President Eisenhower addressed the UN General Assembly, declaring that it was time to create a new international body under the UN, dedicated to the concept of 'Atoms for Peace': the International Atomic Energy Agency.

The approval of a draft Statute for the Agency did not come until 26 October 1956. Knotty questions included the matter of accounting for fissile material, the arrangements by which Agency inspectors were to carry out accounting, and in general the whole inspissated issue of 'safeguards': the guarantees which would assure that 'atoms for peace' stayed that way. We shall have more to say about this critical consideration in Chapters 9 and 10.

Other Agency activities, perhaps less crucial but also less controversial, made much better headway. The first Atoms for Peace conference in Geneva in 1955 was followed by another in 1958, and two later ones, in 1964 and 1971, each more successful than its precursor in exchange of technological information and mutual enthusiasm. The Agency headquarters in Vienna became a major clearing-house for international nuclear activities, and a vigorous proponent of nuclear benefits for all.

In December 1957, under the auspices of the Organization for European Economic Cooperation (OEEC) eighteen European countries joined the European Nuclear Energy Agency, to further the joint development and harmonization of civil applications of nuclear energy. In due course it established three joint projects: the Dragon project (see pp. 59–61), the Halden boiling heavy water reactor project in Norway, and the Eurochemic fuel reprocessing facility at Mol in Belgium. Exchanges of scientific and technical information were promoted, international working groups of experts were set up to consider specialist areas, and efforts began to coordinate nuclear legislation and radiological protection. In 1960 the OEEC became the Organization for Economic Cooperation and Development (OECD) encompassing also the USA, Canada and Japan. When Japan became a full member, the ENEA dropped the 'European', to become the OECD Nuclear Energy Agency.

The first major post-war economic partnership in Europe was the European Coal and Steel Community, whose activities laid the groundwork for the European Economic Community. The countries of the soon-to-be EEC decided to establish a joint enterprise, modelled on the Coal and Steel Community, to be called Euratom. However, Euratom's grandiose programme of research, development and supranational administration of nuclear matters came unstuck when confronted with stubborn national interests.

Instead the international aspect of nuclear activity began on an essentially bilateral basis, between pairs of countries who could see mutual advantage in exchange of information and technology. Much of the initial momentum for such exchanges came, oddly enough, from US interests who considered that other countries with energy shortages might provide a sizeable market. It was in part this aspect which prompted otherwise uneconomical investments in nuclear stations within the USA; foreign buyers, it was felt, would have reservations about purchasing a technology not yet established in the domestic US market.

The USA supported foreign nuclear development by supplying enriched uranium at knockdown prices and by technical assistance, some of which subsequently re-emerged in the form of vigorous indigenous development of pressurized water reactors and boiling water reactors in Sweden, West Germany and Switzerland, so that the leading European firms became stiff competition for the American industry. West Germany was the first non-weapons nation to start up a nuclear power station, a 15-MWe BWR at Grosswelzheim in November 1960. Meanwhile Britain sold two Magnox reactors overseas, to Italy and Japan. The Latina Magnox station went critical in December 1962, and the Tokai Mura Magnox station in May 1965. These were the only British power reactors ever to find foreign buyers. In Sweden the 12-MWe Ågesta power station, using a Swedish design of pressurized heavy water reactor, went critical in July 1963.

Canada and Britain had experienced their first serious reactor accidents at a chasteningly early stage in their development programmes: not so the USA, who managed nearly two decades with only comparatively minor mishaps involving reactors. The EBR-1 Mark II core melt accident was by no means trivial, but it resulted in only minor exposure to personnel. Other accidents included the destruction of the BORAX experimental reactor in 1954, fuel damage to one of the Hanford production reactors, and fuel melting in the Heat Transfer Reactor Experiment, the Sodium Reactor Experiment and the Westinghouse Test Reactor. Damage was in several instances costly, as was clean-up. But all these assorted episodes, while not precisely reassuring in their variety and frequency, were more or less minor.

The first major reactor accident in the USA, when it finally occurred, was not only major but ugly. On 3 January 1961 at 4 p.m. John Byrnes, Richard McKinley, and Richard Legg, three young servicemen, went on duty at Stationary Low-Power Reactor No. 1 (SL-1) at the NRTS in Idaho. SL-1 was a 3-MWt prototype military nuclear power plant. It had been shut down for work on instrumentation, and the control-rod drives disconnected. Byrnes, McKinley and Legg had been detailed to reassemble these drives. This required that the central control rod be lifted just ten centimetres and coupled to the remote driving mechanism, a straightforward procedure which the three had carried out many times. No one knows exactly what happened on 3 January 1961. Later reconstruction of the fatal four seconds indicated that the refit had actually been completed. Then, for reasons that will remain forever unknown – thoughtlessness, horseplay perhaps – the central control rod, number 9, was pulled out of the core. The official report by AEC investigators suggests that the control rod was stuck, and that Legg and Byrnes tried to heave it up manually. When it came loose it rose not merely ten centimetres but nearly fifty. The result was catastrophic. The core almost instantly went supercritical, the fuel fried itself, and the resulting steam explosion blasted a virtually solid slug of water at the roof of the reactor. The reactor vessel rose three metres, right through the pile cap. Legg and McKinley were killed instantly; McKinley's body was impaled in the ceiling structure on an ejected control rod plug. Byrnes was cut down by a withering flash of radiation. Automatic alarm systems brought emergency squads, but even before they reached the reactor their radiation dose-meters were reading off-scale, more than 500 roentgens per hour, a lethal level of radiation. The level inside the reactor building was even higher, more than 800 roentgens per hour. Nonetheless two rescuers rushed into the wreckage and dragged out Byrnes. But Byrnes died in the ambulance on the way to the Idaho Falls hospital.

Recovery of the other two bodies from the reactor room was a protracted and difficult operation, and had to be carried out with remote handling gear. In the emergency recovery operation fourteen other men received radiation doses of more than 5 roentgens, some of them considerably more. All three bodies remained so radioactive that twenty days elapsed before it was safe to handle them for burial; they had to be buried in lead-lined caskets placed in lead-lined vaults. Not for many months did the level of contamination in the SL-1 building fall low enough to permit investigation of what had happened.

Another military reactor may have been involved in a yet more serious disaster two years later – no one will ever know for sure. Vice-Admiral Hyman Rickover, an uncompromising martinet, ruled the AEC Office of Naval Reactors with autocratic zeal. Rickover was one of the most awkward customers that any nuclear supplier had to deal with. His attitude to quality control verged on the fanatic – as is, in the context of nuclear reactors, both right and reasonable. Nonetheless, in April 1963, the USS

Thresher, a nuclear submarine with 112 navy men and 17 civilians aboard, made a deep test dive about 330 kilometres off the coast of Cape Cod, and never returned to the surface. To this day it is uncertain what became of the *Thresher*; in a vessel of such complexity the possible varieties of malfunction are innumerable. Some knowledgeable observers have no doubt that the *Thresher* suffered a reactor accident, citing features of the fragments of debris said to have been recovered. In any event possible reasons for the disappearance were advanced by Admiral Rickover himself, both before and after the event. The Admiral repeatedly lamented recurrent problems with submarine builders and suppliers: management remote from the designers and shop-floor workers, inadequate supervision and inspection, failure to meet deadlines, failure to meet specifications – even to the extent of using the wrong materials – failure to complete jobs properly, failure to discover other failures because of sloppy maintenance checks. Testifying in the 1964 hearings about the loss of the *Thresher* Admiral Rickover revealed that an inspection shortly before her final voyage located so many faulty welds in critical piping systems that it seemed possible that 'the ship had several hundred substandard joints when she last went to sea'. The loss of the *Thresher* underlined the implications of substandard quality control in nuclear systems, both military and civil, a problem which was to become perennial.

From 1954 onwards the AEC was exercising two functions, laid down by the Atomic Energy Acts, which were fundamentally in conflict. On the one hand it was charged with operating its own nuclear facilities and promoting others; on the other hand it was also solely responsible for laying down and enforcing regulations for the safety of personnel and public *vis-à-vis* nuclear energy and radiation. In March 1961, to forestall more comprehensive action from outside, the AEC arbitrarily divided itself into two. One section was to cover operating and promotional functions, the other licensing and regulatory functions. Although this system operated for thirteen years it came increasingly under fire. The nominal separation of promotional and regulatory activities left many onlookers unconvinced, especially after the surge of private nuclear activities that began in 1963.

In December 1963, New Jersey Central Power & Light Company announced that it was ordering from General Electric a boiling water reactor of more than 500 MWe – more than twice the size of any previous nuclear plant; this plant, to be built at Oyster Creek, would be built without federal assistance. It seemed that nuclear power had at last made the economic breakthrough so long prophesied. The Oyster Creek order began a trickle that was to become a flood.

Southern California Edison ordered the 430-MWe San Onofre pressurized water reactor, which became – in March 1964 – the first of the new generation of plants to receive a construction licence. Connecticut Yankee Atomic Power ordered the Haddam Neck PWR; Commonwealth Edison added a second BWR to its Dresden station; Niagara Mohawk Power Corporation ordered the Nine Mile Point BWR; Rochester Gas & Electric ordered the Robert Emmet Ginna PWR; Connecticut Light & Power ordered the Millstone BWR; Consumers Power of Michigan ordered the Palisades PWR. Commonwealth Edison, by now on the way to becoming the most nuclear-powered utility in the country, added a third BWR, Dresden 3, twin to Dresden 2, and followed with a whole new station not far away, Quad Cities 1 and 2. By mid 1965 the upturn in ordering of nuclear stations in the USA was accelerating remarkably. By the time the San Onofre station had been run up to full power, in January 1968, orders for nuclear stations totalled a capacity of nearly 50 000 MWe and the rush was if anything growing more hectic; another 22 000 MWe were ordered in 1968.

Not all these orders met a welcome in the neighbourhoods expected to play host. From 1961 onwards stirrings of reluctance, foreshadowed by the Enrico Fermi 1 opposition, began to emerge elsewhere. On 12 June 1961 Supreme Court Justice Brennan rendered the Court's majority verdict in favour of granting the Power Reactor Development Corporation a licence to construct the Enrico Fermi 1 fast reactor station; the decision was by no means unanimous, with Justices Douglas and Black contributing a blistering dissent. But the AEC and the Power Reactor Development Corporation got their go-ahead, with consequences we shall describe shortly. Meanwhile, in the summer of 1961, another confrontation became nuclear. For more than two years the Pacific Gas & Electric Company (PG&E) had been quietly acquiring land at Bodega Head, about eighty kilometres north of San Francisco. Local opposition, already stubborn, became yet more vehement when in mid 1961 PG&E announced that the proposed power station was to be nuclear. Until this point the major focus of objection had been wilderness conservation, but now a quite different dimension entered the debate: the site was only 300 metres from the notorious San Andreas fault, the most prolific source of earthquakes in the country, its record including the 1906 earthquake which almost wiped out San Francisco. The idea of situating a nuclear reactor, with its awesome inventory of radioactivity, atop a fault zone seemed to many people an example of foolhardiness verging on lunacy.

PG&E insisted that their reactor and its containment would be sufficiently strong to withstand any conceivable tremor. By 1963 the utility had begun to excavate foundations, although the local opposition, led by a Sierra Club member and lawyer named David Pesonen, were pursuing the matter hotly through the licensing process. By October 1963 two surveyors from the US Geological Survey were on the scene, examining a newly exposed fault cutting across the actual excavation itself. But PG&E stuck to their assertion that the site was suitable, and that their design would overcome any obstacles. Then, as the argument grew ever more heated, PG&E announced on 30 October 1964 that it was abandoning its plans, saying that they 'would be the last to desire to build a plant with any substantial doubt existing as to public safety'. A decade later the hole in Bodega Head remains as a monument to the confidence of 1963 that became the doubt of 1964.

Consolidated Edison likewise shelved plans for a nuclear

station at Ravenswood, in Queens, New York City; the plans, announced on 10 December 1962, met fierce opposition, and provoked David Lilienthal, the first Chairman of the AEC, to declare that he 'would not dream of living in Queens if a huge nuclear plant were located there'. In early 1964 Consolidated Edison changed their minds – at least *pro tem*. On the other coast Pacific Gas & Electric's plan for a nuclear station on the famous beach at Malibu was withdrawn after a wrathful local outcry.

The Enrico Fermi 1 plant, however, duly took shape south of Detroit, and went critical in August 1963. Thereafter, a succession of problems kept it far below its design rating, when it was not actually shut down. Fuel swelling and distortion, sodium corrosion in the core, problems with fuel handling gear, and endless trouble with the steam generators pushed costs sky-high and kept electrical output to a trickle. At last, however the obstreperousness in the steam generators seemed to have been overcome; operators prepared to start up the reactor. On 4 October 1966 the control rods were inched out of the core, and the temperature of the sodium coolant began to rise. The reactor was kept at a low power level overnight; the following morning, 5 October, power-raising commenced. A valve malfunction occupied the morning; from lunchtime until 3 p.m. the power level was brought up to 20 MWt, with another interruption to deal with a malfunctioning pump. Just before 3 p.m. the reactor operator noticed a neutron monitor sending erratic signals from the core. He switched over from automatic to manual control. When the erratic signals ceased, power raising recommenced. Five minutes later, at 34 MWt, the abnormality showed up again. Other instruments appeared to indicate control rods withdrawn more than normal, and unusually high temperatures at two points in the core. But before the control room staff could work out what was happening radiation alarms began to sound.

At 3.20 p.m. six scram rods were inserted to shut down the reactor, and plant staff found out where the radiation was coming from, and why the reactor was behaving oddly. Samples of the sodium coolant and the argon cover gas were discovered to be laden with highly active fission products. Clearly – for reasons utterly obscure – part of the fuel in the core had melted.

The implications of this situation were ominous in the extreme. The Fermi reactor was a fast breeder: its core was a compact cylinder only about 75 centimetres high and 75 centimetres in diameter, the size of a bass drum, which was designed to be capable of producing more than 200 MWt – that is, as much as 200 000 one-bar electric fires. To achieve this awesome output its fine fuel pins, 14 700 of them, made of 28 per cent enriched uranium clad in stainless steel, had to be aligned to meticulous tolerances, no more than a millimetre or so apart. Furthermore this configuration had to be maintained at a temperature of over 400°C, while submerged in an upflowing torrent of liquid sodium passing through the miniscule channels between the pins. Any disturbance of the Fermi core geometry could impede the flow of coolant, leading to unbalanced thermal expansions and more distortion.

The core geometry had another crucial characteristic, common to fast breeders. Unlike a thermal reactor, whose fuel is usually arranged in an optimum geometry to maximize reactivity, a fast reactor has its fuel in a configuration which may be considerably short of the maximum theoretical reactivity it can exhibit. If the Fermi core had been distorted and melted, it might thereafter be susceptible to local surges of reactivity, intense hot spots, which could lead in turn to chemical reactions between fuel, cladding and coolant, and even to violent chemical explosions. Such chemical explosions, rebounding in a collapsing mass of highly enriched fissile fuel, might even cause a full-fledged nuclear explosion.

No one at the Fermi plant had any very persuasive idea of what to do. Any attempt to enter the reactor with the usual remote-handling gear might disturb the precarious equilibrium in the ruptured core. One blunt absolute loomed out of the nerve-racking uncertainty: they had better not do anything hasty. By one of those ironies which in nuclear history seem to abound, the appalling outcome some citizens of Detroit had warned about all the way to the Supreme Court had come within an ace of occurring.

When the full dimensions of the accident became clear an alert went out to all local police and civil defence authorities to prepare *for emergency evacuation of Detroit and other centres. So, at least, insist some of those who received the alert. Official records now show no evidence of an alert; the only way to reconcile the conflicting stories is to presume that a directive was subsequently issued to expunge all trace of it. Be that as it may, the staff at the Fermi plant let some weeks elapse before their first gingerly venture to investigate the core. Eventually, after nearly a year, they ascertained, through a viewing probe specially made to deal with the opacity of the sodium, that something was adrift in the bottom of the reactor. When the Fermi engineers found out what it was they were furious.*

Some years previously, while the Advisory Committee on Reactor Safeguards were still reluctantly mulling over the Detroit Edison proposal, the Committee had devoted much time to the possibility of a core meltdown in such a fast reactor. In particular they feared that a meltdown might allow the concentrated fissile material to collapse into a fast critical assembly. Such a rapid addition of reactivity and the resulting energy release might blow apart the reactor containment and strew its radioactivity over the landscape. This possibility nagged so persistently that the ACRS – when the reactor was in the final stages of construction – demanded a special safety precaution to prevent the collapsing mass of fuel from settling in a concentrated mass in the central volume of the containment under the core. They directed that a metal pyramid be built on the floor of the containment, so that molten fuel would run off its sides and spread out. The engineers constructing the reactor protested vehemently at being compelled to add this odd bit as an afterthought at the last minute. As it turned out, they were vindicated. One of the six zirconium triangles which formed the cover of the 'core-catcher' pyramid was not securely anchored. At some stage the uprush of liquid sodium lifted this zirconium triangle, about twenty centimetres long, and swept it into the precisely aligned core, partially blocking the sodium flow. The temperature of the inadequately cooled fuel pins soared; they melted, warping other pins, further occluding the coolant passages, in a progressive distortion of the assembly which – fortunately – stopped short of a complete meltdown.

Had it not stopped short, no one is quite sure what would have *been the outcome. One possibility was sardonically labelled the 'China syndrome'. The molten mass of highly reactive fuel, generating its own fierce heat and far beyond any hope of cooling or control, might sear its way through all containments and into the rock below the foundations of the reactor, melting, burning and exploding as it went, bound for China. The concomitant outpouring of radioactivity, assuming the accident had opened a pathway to the surroundings, would make the locality a no man's land indefinitely.*

It did not, however, happen. Not quite.

In Britain, all four reactors of the Berkeley and Bradwell Magnox stations were up to full power by the latter half of 1962, and on their way to becoming mainstays of the CEGB system. The Trawsfynydd station barely beat Hinkley Point A to full power in early 1965; Dungeness A and Sizewell A followed in early 1966. The Oldbury station, ordered in early 1961, was half-finished by early 1965, roughly on schedule; the giant Wylfa station, ordered in mid 1963, was making slightly slower headway. North of the border the one nuclear station in Scotland, Hunterston A, had reached full power in late 1964. The nuclear contribution to the British electricity supply by this time far outstripped that anywhere else in the world, including the USA.

But the Magnox design seemed to have come to the end of its role; the capital cost associated with the enormous reactors of Wylfa was unappetizing, and more compact designs, using enriched uranium, now looked more promising: in particular the American pressurized water reactor and boiling water reactor, and the British advanced gas-cooled reactor, of which at that stage only the 32-MWe Windscale prototype existed. By the end of 1964 the CEGB was readying itself to order a second generation of nuclear stations, to follow the Magnox series.

Three separate industrial groups were prepared to tender; two offered versions of the AGR, the PWR in partnership with Westinghouse, or the BWR in partnership with US General Electric respectively. The third group, led by Atomic Power

Constructions offered an AGR design closest to that of the UKAEA, and – by pushing tolerances to the limit – were able to persuade the CEGB and the government that their tender was likely to be most favourable. On 25 May 1965 the Minister of Power told the House of Commons that the second nuclear power programme would be based on the AGR design. In the summer of 1965 Atomic Power Constructions won the contract to build Dungeness B, the CEGB's first twin-reactor AGR station, next to its reliable Dungeness A Magnox station.

The consequences were traumatic for the British nuclear industry. Atomic Power Constructions proved unable to cope with either management problems or acute engineering difficulties, and in due course folded. Contracts for the Hinkley Point B and Hunterston B twin-reactor AGR stations went to The Nuclear Power Group, one of the other two British consortia; British Nuclear Design and Construction received contracts to build the Hartlepool and Heysham stations. The contracts were all 'turnkey' contracts, by which the constructors fixed a price for the entire station, at which it was to be turned over to the utility ready to run. As this is written none of the five AGR stations has yet gone critical, much less run up to power. Dungeness B is expected to take more than ten years to complete, and none of the stations is expected to be able to operate at its full designed power output. Britain's impressive head start into civil nuclear power foundered so badly on the AGRs that it has never recovered. The final outcome is still unresolved; we shall describe subsequent stages in Chapter 8 (pp. 200–202).

In Sweden, the 10-MWe Ågesta pressurized heavy water reactor, designed and built by the Swedish firm of ASEA, went critical in mid 1963 and reached full power in early 1964. It was intended as a forerunner to a series of Swedish-designed power reactors of a heavy water design; but the first full-scale successor, the Marviken reactor at Norrkoping, never attained criticality. Successive modifications failed to overcome unfavourable core characteristics, which eventually compelled the builders to abandon the project. A fossil-fuelled steam plant was built to power the turbogenerator, prompting locals to dub the Marviken plant 'the world's only oil-fired nuclear station'. Fortunately for ASEA, they had already developed an indigenous boiling water reactor; the Oskarshamn-1 station was ordered in 1965, began construction in 1966, and went critical in late 1970.

Switzerland, too, took its first steps towards nuclear power with an indigenous design, with similarly unhappy results. The Lucens reactor, the first Swiss power reactor, was an experimental design, a pressure-tube reactor cooled by carbon dioxide and moderated by a tank of heavy water. By a stroke of what proved to be remarkable foresight it was built in a cavern under a hill.

The Lucens reactor went critical in December 1966; it lasted just over two years. After first reaching its nominal power of 30 MWt on 9 September 1968 it operated at about 7.5 MWe until 24 October, when it was shut down as planned for routine work. The reactor was started up again on 21 January 1969. Power raising began about 4 p.m. At 5.20, with the reactor at a power level of 12 MWt, radiation monitors and pressure-change detectors scrammed the reactor and shut the ventilation valves as the reactor cavern filled with radioactivity. Fortunately the cavern was isolated by air locks; even more fortunately no one was on the reactor side of the airlocks when the alarms sounded. Instruments in the control room indicated radiation dose-rates of hundreds of rem per hour in the reactor cavern. It was apparent in the control room that the coolant pressure, normally fifty atmospheres, was dropping rapidly; within ten minutes it was down to the equilibrium pressure in the reactor cavern, just slightly over atmospheric pressure. The carbon dioxide coolant from the primary circuit had somehow escaped more or less completely into the reactor cavern: the reactor had suffered a loss-of-coolant accident. However, there was still carbon dioxide coolant, albeit no longer pressurized, inside the plumbing of the reactor. The cavern, with its roof of rock fifty metres thick, had – not surprisingly – withstood the pressure shock produced by the bursting primary circuit. Instruments indicated steadily falling temperatures, showing that the primary gas circulators were able, despite the loss of pressure, to remove sufficient heat to prevent the radioactivity of the accumulated fission products overheating the core. Since the Lucens reactor was gas-cooled, its comparatively low power density and the residual coolant left in the primary circuit made removing 'decay heat' from fission products manageable. As we shall see shortly, the consequences of a loss-of-coolant accident in some other designs might not be so readily controlled.

Calls went out immediately to Swiss government nuclear safety authorities. Radioactivity was found to be leaking slowly from the sealed reactor cavern into the adjoining machine hall and fuel store; ventilation for these chambers was sealed. The radioactivity made its way along the subterranean gallery toward the control room; at about 6.15 p.m. control staff were instructed to put on gas masks. The steady leakage of radioactivity had thus far been confined within the facilities buried in the hillside, but it was clear that further leakage might produce radioactive contamination outside the plant. Radiological safety teams were called in, and began to monitor the activity levels on the hillside outside the plant. Monitoring continued throughout the night; no undue activity was located, except from rubidium-88, with a half-life of only eighteen minutes. Radioactive contamination inside the plant was found to be primarily rubidium-88, and iodine isotopes.

By late afternoon of the following day the levels of radioactivity in all the underground chambers except the reactor cavern itself had subsided considerably. The control room, the access gallery, the fuel store and the machine hall were once again coupled to ventilation systems leading to the outside air through filters. Monitoring teams continued to measure activity levels around the hillside, and were able to confirm that no detectable activity was being released. Plant staff were able to return to the machine hall and the fuel store by the following morning, to get ready to enter the reactor cavern itself. The radioactivity in the cavern could be sampled from the machine hall, and included rubidium-88, isotopes of iodine and tritium – the last presumably from evaporating heavy water. On the night of 24 January the reactor cavern was ventilated through filters, once again with monitoring teams keeping watch outside. The spilled heavy water was recovered by remote control.

When plant staff could enter the cavern, they found that the fuel elements in the reactor could not be removed by the refuelling machine. Using a long periscope, the investigators ascertained that the accident had originated in one pressure tube, at the outer edge of the core. Failure of the tube itself or of the fuel element inside had burst it catastrophically. The consequent pressure shock had been transmitted through the surrounding heavy water to all the other calandria tubes, crushing almost all of them on to the pressure tubes inside them and jamming the fuel elements immovably inside the pressure tubes. The calandria tank itself also sustained serious damage, allowing its inventory of heavy water to pour out into the refuelling chamber below the reactor. Fission products released in the ruptured fuel channel had been carried all through the primary coolant circuit, as well as throughout the calandria tank and out through its fractures into the reactor cavern.

Decontamination was a protracted and troublesome affair, as was gaining physical access to the core of the reactor. Much of the pile cap of the reactor was hewed away to get at the constipated fuel channels; caution had to be exercised at every stage not only because of lingering contamination but also to avoid fresh contamination, for instance by breaching the cladding on elements while jockeying them free. Access to the core was not in fact gained until 23 September 1970, twenty-one months after the accident. Special equipment had to be designed and built for the operations.

In the final stages the contaminated coolant piping and even the contaminated upper axial shielding had to be removed; the latter was stored in a steel container, specially fabricated, weighing over sixty tonnes. When the clean-up had been completed the mangled remains of the reactor – moderator tank, pressure tubes et cetera – were stored outside the caverns, and it was decided to use the cavern for storage of radioactive waste. In early 1973 the Swiss Association for Atomic Energy, in their newsletter, remarked airily 'It had in any case been the intention to shut down the plant some time in 1970 or 1971. The incident can therefore be considered as an additional and very instructive experiment.' For completeness they *might well have appended two more adjectives: 'expensive' and 'dangerous'.*

By the end of the 1960s there had been major reactor accidents in Canada, Britain, the USA, and Switzerland. The reactor types involved had included gas-cooled, water-cooled and sodium-cooled; graphite-moderated, light-water moderated, heavy-water moderated, and unmoderated; unpressurized, pressure-vessel and pressure-tube; experimental, plutonium-producing and power-producing: combinations of virtually all the major varieties of reactor design. The industry stressed that there had never been an accident in a commercial reactor which had resulted in danger to the public. The Windscale Number One reactor fire certainly posed a threat of danger to the public; but it was not, of course, a commercial reactor. The SL-1 accident killed three men; but they were not, of course, members of the public. The Enrico Fermi accident released no radio-activity off-site – not quite. The logic of the industry argument was impeccable. But by the beginning of 1970 the public was nevertheless beginning to show signs of widespread disquiet. Ironically, albeit perhaps obviously, the reactor designs which met with the most concerted opposition were the light water reactors – pressurized water reactors and boiling water reactors – which had just begun to dominate the world market.

7. The Charge of the Light Brigade

For the first decade of their existence the light water reactors – pressurized water reactors (PWRs) and boiling water reactors (BWRs) – were only two among the many design variations emerging from industry drawing boards. At the end of the 1960s they were edging into a lead that rapidly became virtually insurmountable. Italy and Japan had begun with Magnox stations imported from Britain; Sweden and Switzerland had begun with indigenous designs that proved unsuccessful. By 1970, however, all four nations had changed course definitively in favour of light water reactors, either imported or home-grown. In 1970 France did likewise. Despite a first generation of gas–graphite reactors second in size only to that of Britain, France dropped the gas-cooled lineage flat. The 1970 order for Fessenheim 1, a 930-MWe PWR, opened the door for a veritable deluge of water reactors. From that time onward light water reactors spread over the world so fast as almost to swamp the other two main lines of development, the British gas-cooled designs and the Canadian heavy water designs. But the light water reactors' surging popularity with the industry was closely paralleled by their burgeoning unpopularity otherwise.

It was certainly true, as the industry insisted, that there had never been a major accident in a light water reactor. But there had been some near misses.

The Dresden 2 BWR, near Chicago, went critical for the first time in January 1970. On 5 June, as it was being worked up gradually to full power, an electrical device in the pressure control system sent out a spurious signal, which caused steam line valves to open in such a way that the turbine cut out and the reactor scrammed. This shut off the fission reaction but not of course the decay heat from the fission products in the core. Since the reactor was still in its com-

missioning phase the heat output from the fission products was still at a comparatively low level. In view of what happened for the next eighty minutes this was just as well.

The reactor scram, in cutting down the heat output from the fuel, reduced the boiling, and the level of water in the reactor pressure vessel dropped as bubbles collapsed. The feedwater pumps, after a false start, set to with a will; since the steam line valves were open the pumps met little resistance, and were able to squirt water into the pressure vessel so fast that the water level shot up. Valves closed; the water level dropped again, then shot up again. A level-recorder jammed at the 'low' position; the reactor operator overlooked others reading correctly, and switched the feed pumps to manual control and maximum input. He was so successful that the surging water got almost to the top of the pressure vessel, feeding not steam but water into the steam lines. When the operator, thirty seconds later, tapped the stuck recorder, its needle swung off scale; the operator set feedwater flow at minimum, but a valve leaked. Two out of the three sets of valves he then tried to open failed to do so; one of the failures was not discovered for some minutes.

After more yo-yos of the water in the reactor a fresh surge of steam entered the steam lines, caught up with the water and fired it like a slug along the main steam line. The surge opened one safety valve; the blast of water and steam from this open valve struck the levers of two other safety valves and froze them open, so that live steam continued to roar out of the pressure system into the 'drywell', the space between the reactor vessel and the concrete primary containment. The pressure in the drywell started to rise.

The drywell pressure should have triggered emergency cooling systems. But the high-pressure emergency cooling system was shut down for repairs following earlier damage; in any case, like the low-pressure emergency system, it would not have functioned because the rising pressure in the drywell was not accompanied by a loss of pressure inside the reactor – the conditions under which emergency cooling would be triggered. When it passed 5 pounds per square inch it put the only available drywell pressure gauge off-scale. From then on the reactor staff had no information about the drywell pressure.

Written emergency procedures demanded that if drywell pressure rose beyond 2 pounds per square inch sprays should be turned on to condense the steam accumulating in the drywell. But two licensed senior reactor operators, who had no way of knowing why the drywell pressure was rising, decided nonetheless to ignore the written emergency procedures and refrain from turning on the containment sprays, because the thermal shock of the sudden cold sprays might cause further damage. Instead, thirty minutes after the electrical fault had started the whole merry-go-round, the operators opened valves allowing the pressure in the drywell to escape to the outside atmosphere via the stand-by gas treatment system. The temperature and the pressure of the vented steam were probably well above the design specifications for the gas treatment system, but a pen recorder ran out of paper and the record was lost. Eighty minutes after the initial spurious signal the operators once again went directly against the terms of the AEC operating license and plant procedures; with the drywell pressure still off-scale, they connected the drywell into a cooling system and into drain-pumps to the radioactive waste-water system. Two hours after the spurious signal the reactor was at last brought under control.

The Dresden 2 incident ended happily, and could reasonably be called an incident rather than an accident. But it was uncomfortably close. The water level rose and fell in the pressure vessel so far that the core was probably left partially exposed at least once. Commonwealth Edison, owners of the plant, later denied that fuel had been damaged; but the Dresden 2 was refuelled after the incident, although it had been refuelled only two months before. Samples taken in the drywell indicated that at least some fission products had escaped from the core into the primary coolant and thence through the jammed valves into the drywell. The amount of iodine-131 in the drywell was some hundred times the permissible concentration on 5 June, and 82 times on 6 June. The stand-by gas treatment system managed to remove almost all the radioactivity when the drywell was vented, even though the pressure and temperature – which the operators did not know when they vented the drywell – were well outside normal working limits.

As an example of the interaction of minor malfunctions, lack of

foresight in design, sloppy operating procedures and plain in-
competence the Dresden 2 incident ought to have been a classic – at
least so far as Commonwealth Edison B W R personnel were con-
cerned. It seemed almost inconceivable that such a situation could
recur. But the Dresden 3 B W R, a twin of Dresden 2, which went
critical in January 1971, experienced on 8 December 1971 – eleven
months later – a practically parallel incident – water levels rising
and falling, water in the steam lines, and 'blowdown' of steam into
the drywell or primary containment, reaching this time a pressure of
twenty pounds per square inch: one-third of the pressure postulated
to follow a complete loss-of-coolant accident in the reactor.

Such incidents rapidly became part of nuclear folklore in the
USA. They resurfaced repeatedly in the courts, especially
following a startling legislative innovation which came into
force on 1 January 1970: the National Environmental Policy
Act, or NEPA for short. The key feature of NEPA was its
requirement that any major development project file with the
newly-constituted Environmental Protection Agency an 'Environmental Impact Statement' (EIS), which would identify
all anticipated environmental effects of the proposed project.
An environmental impact statement was further required to
assess possible alternatives to the proposed project, and otherwise present a credible case *against* the project as well as a case
in its favour. The courts had a field day ruling on issues arising
out of NEPA; for nuclear critics it proved a godsend, giving
them at last the leverage to pry apart the embrace of mutual
self-interest shared by the AEC and the US nuclear industry.

The first test for NEPA in the nuclear field came soon. The
Baltimore Gas & Electric Company had since 1967 been building
a PWR station at Calvert Cliffs on Chesapeake Bay. Objectors,
under the banner of the Chesapeake Environmental Protection
Association, intervened in the hearings for a construction
license. Their opposition to radioactive discharges from the
plant were rejected because the AEC had already set standards
of compliance for radioisotopes like tritium; and the objectors'
concern about the effects of discharge of heated water from the

turbine condensers were declared by the AEC to be outside AEC responsibility, and therefore inadmissible. Similar disclaimers arose in other plant-licensing challenges. The Maryland objectors were not satisfied. They mounted a courtroom challenge to the validity of the AEC's licensing regulations as applied to Calvert Cliffs, in the light of NEPA. The AEC was still stonewalling in late November 1970, and the objectors filed a petition with the US Court of Appeals District of Columbia Circuit, asking for a review of the AEC's refusal to respond.

At this the AEC finally made a move, publishing its extremely narrow interpretation of its responsibilities under NEPA: AEC hearings need not consider environmental factors unless outside objectors raised them; no objections on grounds other than radiological would be countenanced at hearings announced before 4 March 1971; hearings would be prohibited from making independent evaluation of environmental factors if these had already been certified as satisfactory by other state, regional or federal agencies; and no facility already licensed for construction would have to carry out modifications in the light of NEPA.

The Calvert Cliffs objectors renewed their legal challenge. Judgement was delivered on 23 July 1971 by Judge James Skelly Wright. It was the first legal demonstration of the Court's view of NEPA, and it set a stunning precedent. Judge Wright delivered a stinging reprimand to the AEC, declaring that NEPA was not a vague testament of pious generalities but an unambiguous demand for a reordering of priorities in specific decision-making procedures – including, very particularly, those of the AEC. The Calvert Cliffs judgement was a landmark in US judicial history – and indeed in environmental action. Furthermore, it gave a monumental boost to the morale of nuclear opposition groups in the USA and – because of the publicity it received – encouraged the gradually developing liaison between different local groups.

At about the same time, also in the USA, an obscure technical disagreement about reactor engineering had begun to surface, and come to the notice of critical citizens. It is doubtful whether those on either side could have imagined the eventual ramifications of the issue. The disagreement centred on certain auxiliary systems on the two main American reactor types, the PWR and the BWR. The auxiliary systems in question were the emergency core cooling systems (ECCS). A present-day light water reactor of either kind has a comparatively high power density (see pp. 81–2). Suppose a pipe in the primary cooling circuit breaks. The high pressure inside the circuit (up to 150 atmospheres in a PWR) blows almost all the cooling water out through the break very quickly, as water between the fuel pins in the core turns to steam: a 'blowdown', or 'loss-of-coolant accident'. Steam has a much lower heat capacity than water and is much less effective in removing the heat which the core continues to generate. Even if automatic shutdown systems immediately scram the reactor, and stop the fission reaction, the heat output from the accumulated fission products – the so-called 'decay heat' – may be, in the case of a large reactor, well over 200 MWt; this heat output cannot be shut down. Unless some means is provided to remove the heat, the temperature in the reactor core will shoot up with extreme rapidity, until the cladding and fuel softens and melts, and the core begins to collapse. If this happens the consequences – as we have already indicated in the case of the Fermi reactor – may be a catastrophic release of radioactivity to the surroundings. Accordingly light water reactors are provided with a variety of emergency core cooling systems (ECCS) designed to operate automatically – and very swiftly – if a light water reactor's primary cooling circuit is depressurized.

All external connections to a PWR pressure vessel are made above the level of the core. A present-day PWR has three ECCS, one 'passive' and two 'active'. The passive system is an accumulator injection system: two or more large tanks above the reactor, connected into the primary piping, and filled with cool borated water under a pressure of between 13 and 43 atmospheres. If the primary circuit is depressurized, the drop in pressure opens valves and the cool water pours into the reactor. The two active systems are a low-pressure system which supplies replacement water if a large break drops the primary pressure drastically, and a high-pressure system which supplies replacement water if a small break leaves the primary pressure high. Both the high-pressure injection system and the low-pressure injection system involve power-operated pumps and valves, which are activated by monitoring instruments responding to abnormal pressures or levels in the cooling circuits.

A BWR has a drywell containment leading down to a pressure-suppression pool half-full of cool water. Early BWRs have a high-pressure core spray system, later ones a high-pressure coolant injection system, which are activated by low water level in the reactor vessel, by low pressure in the primary circuit, or by high pressure in the drywell (which indicates escape of steam from the primary circuit). If high-pressure injection and the feedwater pumps cannot keep the reactor vessel sufficiently full of water it is fully depressurized, by discharging steam into the suppression pool; a low-pressure core spray then comes into service to spray water from above the core, and a low-pressure reflooding system fills up the reactor vessel from below.

So far so good – if these various emergency core cooling systems do indeed perform as designed. However, by early 1971 some serious doubts were being uttered, particularly by a group of scientists and engineers based in the Boston area. This group, the Union of Concerned Scientists, in mid 1971 first drew attention to a curious and slightly unsettling circumstance whose significance had not previously received much public attention: the total absence of convincing experimental data on ECCS performance. As was subsequently revealed, this absence had been noted with mounting unease by some of the AEC's own staff.

The ECCS saga dates back to at least the mid 1960s, when an AEC task force carried out a survey of emergency cooling. As this is written the saga continues to unfold, and to do more than skim through it would require an entire library. One major document, WASH-1400, weighs over ten kilograms; another, AEC docket RM-50-1, is 22 000 pages long. It must here suffice to note some key features of the controversy. In 1966, at the National Reactor Testing Station, work commenced on the Loss-of-Fluid-Test (LOFT) facility, a full-scale reactor whose original *raison d'être* was as a sacrifice on behalf of ECCS data. The LOFT reactor was to be put through a series of tests, culminating in running it up to full power and so to speak pulling the plug – allowing the coolant to drain away and the reactor to write itself off, with elaborate instrumentation to record its final agonies – inside a massive protective containment, since its death throes were expected to be fairly spectacular.

However, the LOFT reactor ran into a series of delays and cost overruns which eventually made it so expensive that no one could bear the thought of allowing it to destroy itself. It has not yet been completed, nearly ten years after its construction commenced. The design of emergency cooling systems on present-day light water reactors depends almost entirely on computer simulations, based on very little experimental data, and that of questionable quality. Some of the data was generated by Full-Length Emergency Cooling Heat Transfer experiments, designated PWR-FLECHT and BWR-FLECHT, on electrically-heated full-size dummy fuel elements. The PWR-FLECHT tests were carried out by Westinghouse, the main vendors of PWRs, and the BWR-FLECHT tests by General Electric, the main vendors of BWRs. The records of the tests are not reassuring: heaters burned out, thermocouples failed to function, and the Idaho Nuclear Corporation, the AEC contractors then running the National Reactor Testing Station, wrote a series of disgruntled memos pointing out in uncompromising terms how futile the tests were. Another series of tests carried out between November 1970 and March 1971 on a table-top model of a PWR core failed in six trials out of six to get emergency cooling water into the model core after a simulated pipe break.

Nevertheless, in June 1971 the AEC published its Interim

Acceptance Criteria for ECCS. To no one's surprise, almost all reactors then operating or about to be licensed met these 'new' criteria. Shortly thereafter the Union of Concerned Scientists published the first of two memoranda on ECCS (the second appeared in October) taking strong issue with the AEC's expressed policy and drawing attention to the failed model tests. Their warnings about ECCS added fresh fuel to the brushfires of public intervention in nuclear plant licensing processes at sites all over the USA. The AEC, realizing that the ECCS issue would pose intractable problems for its licensing boards, took a drastic step. It announced that it would hold special rule-making hearings on ECCS; accordingly the ECCS issue was thenceforth *sub judice*, and could no longer be introduced into individual licensing hearings.

The ECCS rule-making hearings convened in January 1972, in Bethesda, Maryland, and lasted, with interruptions, well over a year. They generated a record, AEC Docket RM-50-1, 22 000 pages long, plus exhibits even longer, which even by mid summer 1972 was being wheeled about the hearing room on a heavy-duty dolly. The transcript includes some startling material. It chronicles the representations made by various divisions of the AEC itself; by the four vendors of light water reactors – Westinghouse, Combustion Engineering, and Babcock & Wilcox (PWRs) and General Electric (BWRs); by the consolidated electrical utilities; and by the Consolidated National Intervenors, a coalition of more than sixty nuclear opposition groups from all over the USA, who backed the technical testimony of the Union of Concerned Scientists, in the persons of Dr Henry Kendall, a nuclear physicist from MIT, and Daniel Ford, a Harvard economist.

The Bethesda hearings revealed a deep split within the AEC as to the adequacy of ECCS as currently understood. Senior safety experts within the AEC repeatedly went on record as harbouring grave reservations about the effectiveness of ECCS in the event of a loss-of-coolant accident, and about the sketchy basis of data upon which confidence in ECCS was founded. A yet more disconcerting discovery was that senior AEC officials were in the habit of 'censoring' information generated by AEC safety studies if it might prove embarrassing. A series of articles by Robert Gillette in *Science* in September 1972 lent further substance to the accusation that the AEC regularly suppressed in-house information about reactor safety problems which might hamper reactor-builders, and threatened its own employees with the sack if they stepped out of line with the official anodyne AEC view.

Be that as it may, in the spring of 1973 the AEC promulgated the new rules. After all the expenditure of money, time and effort, it was decided that the original AEC acceptance criteria for ECCS were, with minor revision, completely adequate. But the ECCS story did not stop there and the near-misses continued to accumulate.

In the Federal Republic of Germany the first large commercial nuclear power station, the 640-MWe Wuergassen BWR, first went critical on 2 October 1971, supplied its first electricity to the grid on 18 December 1971, and sustained its first accident on 12 April 1972. With the reactor operating at 58 per cent of its design rating a pressure relief valve opened and stuck. To avoid the shock of sudden cooling which a scram would produce, the operators opted for a gradual shutdown. It took about thirty minutes, during which time some 200 tonnes of steam poured from the primary circuit into the surrounding pressure suppression chamber (in US BWRs called the drywell). The steam was condensed in the chamber, but the water temperature rose to 95°C, and pressure surges tore loose structures which had been added to strengthen the floor of the chamber. The water collecting in the chamber then ran out through the seventy-odd screw holes left in the floor, and found its way into the low level of the containment, damaging the electrical wiring of the positioners and motors for the control rod drives. By this time, fortunately, the presence of water in the sump had persuaded the operators to scram the reactor. Lengthy investigations and repairs followed, involving not only the electrical utility but also the Technical Supervision Association, licensing authorities and the Reactor Safety Commission. The Wuergassen plant was recommissioned on 7 November 1972; from mid January 1973 the reactor was run up gradually to 80 per cent of power. From late January it was noticed that water was again accumulating in the pressure suppression system; on 25 February the reactor was shut down to repair an oil leak and to have another look for water leaks. There seemed to be a surprising number of short cracks in one stretch of primary piping, although only one actual leak about twenty-five centimetres long. The investigators suggested that tests carried out before the commissioning of the reactor, especially flushing out of the pipes, had subjected the system to stresses that might have weakened components. The Wuergassen plant was started up afresh in July 1973, but in February 1974 it was yet again shut down for further repairs; as this is written the plant is still out of service.

Nevertheless orders for light water reactors poured in from all sides, not least in the parent USA. After a curious lapse in 1967–8, when the number of operating power reactors in the USA dropped from 22 to 18 – because of shutdowns – the number began to mount, almost all PWRs and BWRs. By 1972, according to the International Atomic Energy Agency, there were 33 operable power reactors in the USA, with an output capacity rated at nearly 15 thousand megawatts. By 1973 this had increased to 56 reactors, with a capacity of over 35 thousand megawatts. The nuclear expansion in the USA was the most dramatic, but the phenomenon was occurring worldwide. By 1973 17 countries had 167 power reactors with a capacity of nearly 61 thousand megawatts. The vast majority were of one or another light water design, giving international flavour to the debate on light water reactor safety which was growing ever more heated within the USA.

On 4 August 1972 the AEC gave a send-off to a major Reactor Safety Study, to be directed by Professor Norman Rasmussen of the Massachusetts Institute of Technology. The $3-million study, funded by the AEC, was carried out by AEC staff and consultants at the AEC offices in Germantown, Maryland, and by AEC laboratories and contractors including

Battelle, Oak Ridge, Brookhaven and Lawrence Livermore. Some of its early results were included in WASH-1250, *The Safety of Nuclear Power Reactors (Light Water Cooled) and Related Facilities*, published by the AEC in July 1973; this was a hefty compendium of technical data and policy considerations which was seized upon by the many organizations by this time embroiled in nuclear confrontations, including Friends of the Earth, the Natural Resources Defense Council, Ralph Nader, the Scientists' Institute for Public Information, the Sierra Club, and a rapidly expanding constituency of local groups.

Objectors were becoming assiduous collectors of nuclear folklore. There was Millstone-1, whose condensers corroded and leaked sea-water into the primary cooling; Quad Cities-2, which operated with a forgotten welding-rig sloshing around inside the pressure vessel; Vermont Yankee, on which the control rods were installed upside down, and which by a combination of ingenious malpractices was later started up with the lid off the pressure vessel; Indian Point-2, in which a major steam pipe split over half its circumference, allowing leaking steam to buckle the steel liner of the containment for more than twelve metres; Palisades, in which the core support barrel worked loose and played hob with the reactor internals, causing an indefinite shutdown and provoking a $300-million lawsuit by the operators, Consumers Power, against the builders, Combustion Engineering; and so on.

Such matters were of at most secondary concern to Britain's Central Electricity Generating Board (CEGB) when in 1973 they casually revealed that they were proposing a new series of power reactors, and that they had opted for PWRs. This revelation, after months of speculation, came in testimony by the CEGB Chairman, Arthur Hawkins, before the Parliamentary Select Committee on Science and Technology on 18 December 1973; and it triggered a furore unparalleled in Britain's long nuclear history.

The CEGB plan called for thirty-two 1300-MWe PWRs to be ordered in the decade from 1974 to 1983, a programme whose magnitude overwhelmed most onlookers, especially as the British nuclear industry was in disorder after the advanced gas-cooled reactors debacle. The government had just agreed to draw together the whole reactor-building industry into one consortium, the National Nuclear Corporation, and to place the management of this consortium in the hands of Britain's private General Electric Company who would also become majority shareholders.

The General Electric Company shared the CEGB's enthusiasm for the switch to light water reactors, but together they drastically overplayed their hand. Before long opponents of the plan included – as well as the British wing of Friends of the Earth – the Parliamentary Select Committee, who published a terse and hostile report in February 1974; the Institution of Professional Civil Servants; other trades unions; many people in the British nuclear industry, including eventually the architect and founder of the British industry, Lord Hinton; Sir Alan Cottrell, a metallurgist of international standing and until April 1974 government Chief Scientist; and many well-informed and influential commentators in the media.

The arguments were manifold. Questions included the safety of light water reactors; public approval of nuclear power; the credibility of the claim that the 1300-MWe design was 'proven' – since none had yet operated anywhere; the effect of imported components on the balance of payments; the accuracy of the forecasts of electricity demand which required such a vast programme; and the effect on British nuclear morale if its twenty years of achievement were abandoned in favour of trans-atlantic technology. The situation was further complicated by the election of February 1974, in which the Heath government – apparently in favour of light water reactors – was superseded by the Wilson government, whose loyalties were different.

Whatever the reasons, and many have been plausibly advanced, the government decision, foreshadowed from June 1974 onwards, was to reject the CEGB's proposals. Instead of PWRs, Britain would build steam-generating heavy water reactors (SGHWRs) – neither so large, nor so many. Authorization was given for just six 660-MWe reactors, four for the

CEGB and two for the South of Scotland Electricity Board (who had favoured this design throughout), with a further review promised in 1978. Despite the government's careful wording, stressing that the decision in no way implied an unfavourable reflection on light water reactors, it was a major setback. Coupled with belated international recognition of Canada's impressive 2000-MWe Pickering CANDU station, the British reactor decision gave a new lease of life to heavy water designs, and made the world predominance of light water reactors seem somewhat less inevitable.

Proponents of light water reactors lost little time in returning to the fray. On 21 August 1974 the draft version of the US Reactor Safety Study was published as WASH-1400, *An Assessment of Accident Risks in US Commercial Nuclear Power Plants*, indicating that such risks were minimal. As published it included a brief summary – in effect a child's guide to reactor safety, in the form of a catechism – plus the main report, plus ten volumes of back-up material, weighing in all something over ten kilograms, and forming a stack about thirty centimetres high.

On the basis of the technique known as event-tree/fault-tree analysis the report concluded that the probability of a serious accident in an American light water reactor nuclear plant was very low indeed – one chance in a million per reactor per year of an accident killing as many as seventy people, and similar low figures for a wide range of other possibilities. In the 'tree' technique a possible malfunction is identified; all the alternative possible consequences of the malfunction are identified; then all the possible consequences of each of these alternatives are identified, and so on. When such a 'tree' has been developed, with more and more branches of possible sequences of effects, probabilities are assigned to each effect, and the cumulative probability of each outcome calculated. This technique was heavily criticized after the publication of WASH-1400; similar techniques employed in the space programme and other high-technology projects had not been conspicuously successful.

There is no doubt that the data and analysis embodied in

WASH-1400 are an impressive dissection of a technological situation of great importance. Whether WASH-1400's impressive dissection gives credible numerical values for the probabilities and consequences of accidents to light water reactors is still a matter for intense debate; obviously a comprehensive critique of such an extensive research study takes time and effort, and very few critics can command resources commensurate with those of the AEC study. In any event the Reactor Safety Study lays down in detail the assumptions underlying the safety philosophy of the US industry, making it possible to establish with much more precision those areas whose analysis can be considered satisfactory, and those still requiring further attention.

Amid all the uproar orders for light water reactors poured in. It seemed for a time that only the other two members of the original Second World War partnership would be able to hold out – Canada with the heavy water CANDU family and Britain with the gas–graphite family. Of course, for power production, Britain had developed the gas–graphite reactors almost by default. Despite the scale of the gas–graphite Magnox and AGR programmes, and the parallel development of the SGHWR, Britain was really looking towards the liquid metal fast breeder reactor (LMFBR).

From the earliest days of the British nuclear effort long-range sights had been set on the LMFBR as the ideal form of nuclear power reactor. The small Dounreay Fast Reactor was followed by the 250-MWe Prototype Fast Reactor, also at Dounreay, ordered in 1966. But problems, especially with the very intricate reactor roof, delayed work on the Prototype Fast Reactor. As a result it was overtaken by its cross-channel rival, the French Phénix fast breeder reactor at Marcoule. The Phénix went critical in August 1973, and reached full power on the final day of a major international conference on fast reactor power stations taking place in London in March 1974, a public relations coup which did not go unremarked. The Prototype Fast Reactor itself went critical just before the opening of the conference.

The USA, too, was by this time in hot pursuit of the

LMFBR, albeit some distance behind the British, French and Soviets. The 16.5-MWe Experimental Breeder Reactor-2, successor to the Experimental Breeder Reactor-1, went critical in the summer of 1963 and was run up very gradually to full power, which it at length attained in mid 1969.

In August 1968 the AEC published its LMFBR Programme plan, ten volumes numbered WASH-1101 to WASH-1110, subsequently revised but unambiguously retaining its boyish enthusiasm. On 7 August 1972 the AEC, the Breeder Reactor Corporation and the Project Management Corporation signed a Memorandum of Understanding for the construction of what the AEC was pleased to call the first prototype fast breeder power station. The occasion carried on the persistent rewriting of history, to the effect that the Enrico Fermi 1 plant had existed only as an experimental facility. Older records reveal, of course, that the Fermi 1 plant was envisioned as fulfilling precisely the role now assigned to the new 'first' prototype LMFBR plant: a working power station to demonstrate the feasibility of the LMFBR design for electricity supply. Since the Enrico Fermi 1 plant had signally demonstrated exactly the contrary it is understandable – albeit deplorable – that it should be swept under the carpet.

In any event, mindful of the National Environmental Policy Act, the AEC had dutifully published in July 1971 a draft environmental impact statement on its LMFBR demonstration plant – well before any detailed design work had been done, and indeed before a site had been chosen. The usefulness of such an abstract statement of environmental impact can be imagined. The AEC published at about the same time an environmental impact statement for the Fast Flux Test Facility, a sodium-cooled experimental reactor then – and in 1975 still – under construction at Hanford; it was to become one of the most flagrant examples of cost escalation that even the nuclear development programme had ever manifested, a worrying omen for the AEC's LMFBR programme. This programme was grandiose even by AEC standards, envisioning some 400 LMFBR power stations in the USA by the year 2000. To this end, the AEC was devoting fully one-quarter of the total Federal funding for all energy research and development to the LMFBR: about $500 million per year by the mid 1970s. All other forms of US Federal energy research and development – not least among them studies of the safety of light water reactors, plus all other research into nuclear fission and fusion, coal technology of all kinds, and needless to say the modicum of funding provided for alternative sources of energy – sun, wind, geothermal and so on – had to share the remainder. None of the shares came even remotely near that allotted to the LMFBR.

In early 1973 the Scientists' Institute for Public Information teamed up with the Natural Resources Defense Council, and filed suit against the AEC over the LMFBR programme. They averred that the National Environmental Policy Act (NEPA) required from the AEC not merely an impact statement for its LMFBR Demonstration Plant, but instead an impact statement for the whole LMFBR programme towards which the Demonstration Plant was only a tentative step. On 12 June 1973 the US Court of Appeals for the District of Columbia agreed with this view, and declared that the AEC must prepare an impact statement for the whole programme. For the first time the overall effect of a new technology would be analysed before its introduction – potentially a historic breakthrough.

In March 1974 the AEC published its draft impact statement on the LMFBR programme: WASH-1535, five volumes containing more than 2000 pages. Very few recognized it as a breakthrough; as a response to the NEPA it was, said the Natural Resources Defense Council and the Scientists' Institute for Public Information, neither serious nor in good faith. The Environmental Protection Agency concurred, calling the draft 'inadequate', and sending it back for total rewriting. The AEC, who had been expecting to wrap up a final approved impact statement by mid 1974, eventually announced a delay which, at this writing, seems indefinite. However, the plans for the Clinch River Breeder Demonstration Plant near Oak Ridge, Tennessee, proceeded apace, even as cost estimates soared from $700 million to $1700 million to over $2000 million. It was perhaps significant that the 340 utilities who collectively agreed to put up $250 million – that is, an average of under $700 000 per utility, scarcely a vote of confidence in a new energy technology – did not increase their contribution. The escalating costs would come out of public funds exclusively.

In early 1974, just before the London fast reactor conference, reports in the French newspaper *Le Monde* and elsewhere suggested that the Soviet BN-350 fast breeder reactor at Shevchenko on the Caspian Sea was also having its troubles. The BN-350 was the first of the new generation of fast reactor power stations to go critical, in November 1972, beating the French Phénix by nine months and the British Prototype Fast Reactor by fifteen months. But the *Le Monde* report, referring to evidence gathered by a US reconnaissance satellite, declared that the BN-350 had suffered an accident of some kind – apparently one whose effects were visible from the outside of the installation.

At the conference, Dr N. V. Kraznoyarov, Deputy Director of the USSR Atomic Energy Research Institute at Dimitrovgrad, revealed in response to pointed questions that – like all its brethren – the BN-350 had had problems with steam generators, in keeping the sodium and the water from fraternizing. There had, he said, been leaks in three of the plant's six steam generators, two minor, and one of about 100 kilograms of water. He denied that there had been an 'explosion' at the BN-350, but made no mention of a fire, leaving many delegates with the impression that a fire had indeed occurred. Certainly a sodium–water reaction involving 100 kilograms of water would have released a substantial amount of energy. Dr Kraznoyarov indicated that no effort had yet been made to enter the steam generator to assess the damage, leading to the suggestion that radioactivity had been released. At the time of the conference, the BN-350 was, according to Soviet spokesmen, again in operation, on its three remaining steam generators, working at 30 per cent capacity. As in other matters of technology the Soviets are not in the habit of discussing their own problems openly, and the London conference did not produce much further detail.

Meanwhile they continued work on the BN-600, another step up in size, at Beloyarsk. The French, satisfied that this intermediate step could be bypassed, prepared themselves, in concert with other European partners, to construct a 1200-plus-MWe Super-Phénix. The British continued their meditations on a commercial fast reactor of similar size. Although the advent of the first commercial fast reactor seemed to recede as time advanced, a policy paper presented by the United Kingdom Atomic Energy Authority (UKAEA) to the March 1974 conference declared that Britain should have twenty-five 1000-MWe fast reactor power stations in operation by the year 2000, a prospect which struck some observers as – to say the least – improbable. Improbable or not, the UKAEA spent some £32 million in the year 1973–4 on fast reactor development – more than two-thirds of its power programme budget.

One sector of the British nuclear industry had a more immediate matter on its mind, a reminder that problems could arise not only in reactors but also elsewhere in the fuel cycle. British Nuclear Fuels, since 1971 operators of Windscale and the other production facilities hived off from the UKAEA, had converted building B204 to serve as a head-end plant to preprocess oxide fuel. But on 26 September 1973, at 10.55 a.m., as workers were starting to feed in a fresh batch of irradiated fuel, radiation alarms began to sound. A health physics monitor, warned about a beta-radiation alert on the tenth floor, passed the word to the plant foreman on the ninth floor. The monitor then continued down the stairs, shouting to the rest of the staff to get out of the building. While the staff made their way to the Health Physics Control Room in Building B217, two senior men with respirators searched building B204, and found four more men who had not heard the warning. The plant foreman and plant manager, also wearing respirators, re-entered building B204 at 11.15 and within half an hour had shut down the plant. All thirty-five personnel who had been in building B204 were found to have skin and lung contamination from the beta-emitting radioisotope ruthenium-106. The skin exposure was not considered to be significant; but quantities inhaled into the lungs of thirty-four of the men ranged up to 5 microcuries, and one man was found to have 40 microcuries in his lungs. The National Radiological Protection Board estimated that this would increase by about 1 per cent the likelihood of his contracting lung cancer during his lifetime.

A painstaking investigation by the Nuclear Installations Inspectorate at length identified the quite unforeseeable confluence of circumstances which had led to the release of the radioactivity. A vessel in the process line was found to have accumulated a layer of solid radioactive residue, which apparently reacted unexpectedly with an influx of acidified Butex solvent from the start-up of the process line. The reaction generated enough gas pressure to expel some of the churned-up radioactivity out through a seal on a drive-shaft entering the vessel chamber. The Inspectorate's report on the episode laid down a series of recommendations which in effect required that this part of the reprocessing plant be hermetically sealed before it was put back into service. The section – the head-end plant for oxide fuel (see pp. 101–2) – was expected to be out of action indefinitely. The Inspectorate also noted with unhappiness the unconcern of the plant staff, who had paid little attention to the radiation alarms. The alarm systems themselves were found inadequate in type and number, and the emergency arrangements left a good deal to be desired. But the report conceded: 'The lack of suitable emergency arrangements was evidently due to the fact that there had been no similar incident on the site or, as far as is known, anywhere else. There appears to have been no reason to expect such an incident.'

Reprocessing was also causing trouble in the USA. Ten years earlier, in 1964, General Electric had come up with a new technique for reprocessing irradiated reactor fuel that did not generate copious and troublesome quantities of high-level liquid wastes. The process was based on the chemistry of the volatile fluorides of uranium and plutonium. General Electric was sure the new process, called Aquafluor, would be a breakthrough in reprocessing, compact and efficient enough to be used on reactor sites themselves. In 1968 General Electric began construction of a plant based on the Aquafluor process, the Midwest Fuel Recovery Plant, near Morris, Illinois, south of Chicago. Six years later, in July 1974, after well over two years of struggle, General Electric filed with the AEC a report which admitted that the Midwest plant did not work, and probably never would. Its cost had almost doubled over the original estimate of $36 million, but General Electric's report offered little hope that much if any would be recovered. The plant is made of massive concrete walls, far from easy to reorganize; but trials revealed virtually insuperable maintenance problems in areas which would be irrevocably contaminated by high-level radioactivity once the plant was operating.

The failure of the plant may have driven General Electric out of the reprocessing business; it undoubtedly left the US nuclear industry in a quandary, with no commercial reprocessing facilities available until 1977 at the very earliest. As a result the backlog of irradiated fuel filled many cooling ponds to capacity, and the AEC had to intervene to make storage room in its own facilities. Meanwhile a steadily increasing amount of the available fissile material, uranium-235 and plutonium-239, was tied up in irradiated fuel, adding to the carrying charges of the nuclear fuel cycle. The Barnwell, South Carolina, reprocessing plant of Allied Gulf, with a capacity of 1500 tonnes per year, may or may not be in service by 1977; the Nuclear Fuel Services plant near Buffalo, NY will be out of action until 1979 for remodelling and enlargement. By then up to 2300 tonnes of irradiated fuel will be awaiting reprocessing. Meanwhile General Electric proposed to make the Midwest plant cooling ponds available for storage of irradiated fuel; since the Midwest design was never intended to safeguard radioactive materials in such a form, leading nuclear critic David Comey of Businessmen for the Public Interest challenged the proposal as being entirely too off-hand.

The Midwest plant was not the only nuclear casualty of 1974. In the summer of 1974 the Kernkraftwerk Niederaibach, a 100-MWe prototype pressure tube heavy water reactor in West

Germany, joined the list of defunct prototypes, only eighteen months after criticality and at a cost of DM 230 million. In Japan the 320-MWe Mihama-1 PWR was facing a long shut-down. It was realized that only a very few leaks in a steam generator might play havoc with emergency cooling in the event of a loss-of-coolant accident: steam might be driven through the leaks into the primary circuit, at a pressure that might keep emergency cooling water from reaching hot spots in time. Mihama-1 had developed so many leaks that there seemed no option but to replace the steam generators, a costly and time-consuming exercise. However, in an awkward year for nuclear power nothing could equal the saga of the *Mutsu*.

The NS Mutsu *was the prototype nuclear cargo vessel of the Japan Nuclear Ship Development Agency. Its voyage of September 1974 probably did more to set back the cause of nuclear marine propulsion than anything in Admiral Rickover's worst nightmares. Nuclear ship propulsion has of course a long history, dating from the* USS Nautilus *of the mid 1950s, including many other nuclear subs, American and otherwise, aircraft carriers and other naval vessels, plus a handful of non-military ventures. None of the latter can exactly be called a triumph. The* NS Savannah, *the first American nuclear cargo ship, found herself a sea-going pariah, unwelcome in almost every port in the world. The* USSR, *perhaps with this situation in mind, built the* Lenin, *a nuclear-powered icebreaker which was not concerned with ports of call, and gained the long-range advantage shared by nuclear subs. West Germany built the* Otto Hahn, *whose career, if unremarkable, has at least been free of scandal. But the* Mutsu *escapade lowered the tone of the argument to the level of the broadest farce.*

The Mutsu, *powered by a PWR, was launched in 1969. She was named after her home port; but her home port was far from fond of her. Local fishermen were deeply suspicious, afraid that radioactive discharges from the* Mutsu *would damage the fisheries, or at any event make it more difficult for them to sell their catches. The* Mutsu *was ready for sea trials in 1972, but public opposition prevented her from sailing. People were worried lest anything go wrong with the start-up of the ship's reactor, with possible deleterious effects on the rich scallop fisheries of Mutsu Bay. Lengthy discussions between local officials and the government led to establishment of a federal fund of 100 million yen to cover compensation if the fisheries suffered; by August 1974 most local officials agreed to let the* Mutsu *sail. But the mayor and some of the fishermen were adamant, and blockaded the nuclear vessel with some 250 small fishing boats. Then, on 25 August, a typhoon forced the fishing boats to run for shelter, and the* Mutsu *slipped out of the bay at midnight under auxiliary power and with a naval escort.*

On the high sea 800 kilometres from the coast, the reactor was brought to criticality on 28 August; but as power was raised slightly for testing a radiation leak was discovered. The leak was apparently minor, but since the reactor commissioning was being done at sea the reactor operators had to improvise. Their improvisations delighted newspaper sub-editors all over the world, and the Mutsu *became overnight a household word, in the most embarrassing possible sense. First the operators tried brewing up borated boiled rice as an impromptu shielding cement; this was mildly successful but not completely; so they resorted to old socks, and the waiting newsmen rejoiced afresh. Clearly it would have been preferable to return to port and carry out the necessary tests and modifications with shore facilities available. But feelings in Mutsu Bay were running so high that the eighty-nine* Mutsu *crew members feared for their safety if they were to attempt to bring the crippled ship into the harbour. So the* Mutsu *floated helplessly off the Japanese coast, while frantic negotiations took place between the stubborn Mutsu fishermen and the nuclear ship developers.*

The ship drifted for forty-five days before it was at last permitted to return, and even then only under the most stringent conditions. The authorities had to agree to find the Mutsu *a new port within six months, to leave the fuel in the reactor, to remove all shore-based nuclear facilities within thirty months, and to turn over to the mayor the keys of the fuel-handling crane; furthermore the government was to provide $4 million, to establish a compensation fund lest rumours of radioactivity injure shellfish sales and to pay for new public works in Mutsu. As an exercise in Japanese nuclear* *public relations the* Mutsu *episode – for far less obvious physical reasons, to be sure – picked up where the* Fukuryu Maru *had left off twenty years before.*

Nuclear public relations in Sweden also took a nose-dive in the latter part of 1974. The debate over nuclear policy in Sweden had been for some time more wide ranging than virtually anywhere else, with full-blooded public participation – eventually, by autumn 1974, to the extent of government funding of some 6000 energy-study groups all over the country. In the Swedish Parliament vigorous speeches had called for careful re-examination of the Swedish nuclear programme in the light of American evidence regarding emergency core cooling systems and other safety measures. Leading American critics were invited to speak in Sweden, often with official support of various kinds. In autumn 1974 Swedish environmental groups including Jordens Vaenner (Swedish Friends of the Earth) and Miljoecentrum commenced a campaign for signatures opposing further nuclear stations in Sweden. They were doing well, collecting names by the thousands, when the Swedish nuclear industry revealed an embarrassing *faux pas*, one which could scarcely have come at a more awkward moment. The Ringhals 2 nuclear station, with an 820-M We Westinghouse pressurized water reactor, was in its commissioning stage when the State Nuclear Power Inspectorate discovered that three high-pressure injection pumps were not only wrongly assembled and installed, but were in fact entirely the wrong pumps, failing to meet even Westinghouse's own criteria. The pumps were ordered to be removed and replaced, and meanwhile the reactor was not permitted to operate at full power. The discovery drew a fresh burst of criticism, all the way up to ministerial level. Further investigation pinpointed two more faulty pumps elsewhere in the plant.

As the catalogue of nuclear headaches lengthened into the 1970s the Canadians alone seemed to have affairs under control. While the Netherlands and even Spain began to encounter the stirrings of discontent, and the roll call of problems elsewhere continued to grow, Canada's huge Pickering station near Toronto loomed as the triumphant show-piece of the nuclear business. By mid 1974, the CANDU reactor, nothing but a curiosity to the international market a year before, suddenly looked a winner. The light water reactors were meeting every kind of difficulty: turned down by the British, challenged by environmental objectors in at least ten other countries, accused of shortcomings in safety, and – perhaps most damaging of all – revealing serious limitations on their performance. Electrical utilities in the USA, in increasing financial straits after the oil price rises, cut back severely on ordering of new generating stations – and the capital-intensive nuclear stations took the brunt of the cutback. By late 1974 eight light water reactor orders had been cancelled outright, and another eighty-six delayed, in some cases for several years. But there was little sign of such reluctance in Canada.

The four 500-MWe CANDU reactors in the Pickering A station had an availability far higher than most other nuclear stations. The capacity factor – total electricity output as a fraction of the maximum possible – was 84 per cent during 1973, with Pickering 1 and 4 both better than 90 per cent. Plans were under way to double Pickering, by adding a B station with four more 500-MWe CANDUs; the even larger Bruce Station on Lake Huron was on schedule; other orders were pouring in. Even the early heavy water troubles seemed to be past. Canada's first attempt to build a heavy water production plant, at Glace Bay, Nova Scotia, in the early 1960s, had come a pricey cropper; the plant did not work, and had to be completely rebuilt at a cost of many million dollars. Canada had been compelled to buy heavy water from the Savannah River installation of the USAEC. But by 1974 the Port Hawkesbury plant in Nova Scotia and the Bruce plant near the Bruce power station were in operation, and even the Glace Bay plant seemed likely to come into service in 1975. It seemed that CANDU could do whatever it wished.

Then, on 10 August, three of the pressure tubes in Pickering 3 were found to be leaking. By the time of the discovery about two tonnes of heavy water coolant had escaped into the space around the cooling circuit. Not only was the coolant radioactive – laden with tritium formed by neutron absorption in the deuterium – it was also, at some $66 per kilogram, too expensive to lose. The reactor was shut down, and a prolonged inspection began. The first three leaking tubes were replaced, but it soon became apparent that similar leaks were developing at many points on one reactor face. The leaks were tiny cracks, the result of incorrect fitting of end-joints, which had exerted too much stress on the new zirconium–niobium alloy used for the pressure tubes. The same alloy was used in Pickering 4, and in the Bruce reactors, as well as elsewhere. As this is written work is progressing on the identification, removal and replacement of leaking tubes in Pickering 3, and a watch is being kept for the same problem in Pickering 4.

However, for thrills and spills it remained difficult to top the light water reactors. After three country-wide pipe-crack shutdowns of boiling water reactors, in September and December 1974, and January 1975, came an episode which won instant nuclear immortality for a humble candle. The candle was in the hand of an electrician in the cable-spreading room under the control room of the Brown's Ferry station in Alabama, which had just become the world's largest operating nuclear station, with two BWRs of nearly 1100 MWe up to power. At 12.30 p.m. on 22 March 1975 the electrician and his mate were checking airflow through wall-penetrations for cables, by holding the candle next to the penetration, when the draught blew the candle flame and ignited the foam plastic packing around the cable-tray. The electricians could not put out the fire. The temperature rise was noticed by the plant operator, who flooded the room with carbon dioxide and extinguished the fire beneath the control room – but the fire had already spread along the cables into the reactor building. When erratic readings began to appear on the controls for Unit One the operator pressed the manual scram button. Soon he found there was also a half-scram on Unit Two, which he had not ordered; and the speed of the main recirculating pump was being reduced. Quickly he scrammed Unit Two.

Until the two units were scrammed they had been supplying some 15 per cent of the total demand on the whole Tennessee Valley Authority grid; the effect of their sudden removal can well be imagined. The fire continued to burn for seven hours, affecting hundreds of cables. According to an initial assessment by the US Nuclear Regulatory Commission – successors to the dismantled AEC – the fire knocked out all five emergency core cooling systems on Unit One. Repairs to the station were expected to keep it out of service for months, and entail over $40 million in the cost of replacement electricity output alone. Nuclear critics and advocates alike agreed that it was potentially the most serious incident in the history of the industry.

By early 1975, after a year of dramatic ups and downs, the nuclear industry world-wide found itself assessing its situation – perhaps, as an industry, for the first time. It was by this time one of the world's largest and fastest-growing industrial sectors. A 1974 list of firms engaged in nuclear business included over 1800 names in the engineering category alone. There were the giants: Westinghouse, General Electric (USA), Shell, Gulf, Exxon, du Pont, Atlantic Richfield, Union Carbide, and many other multinationals – and there were smaller firms, down to those with only a handful of highly specialized employees engaged in some esoteric corner of the technology. There were suppliers of concrete, steel of many different kinds, lead, copper, graphite, boron, zirconium, heavy water, sodium, carbon dioxide, argon, helium, ion-exchange resins, filters, plastics, insulation – the list goes on indefinitely. There were suppliers of complete nuclear power stations, of complete nuclear steam supply systems, of turbogenerator sets and other heavy electrical gear, of pressure vessels, heat exchangers, steam generators, pipes, pumps, valves, instruments, electronics, computers, control rods, cranes, emergency systems, stand-by diesels – and on and on. Besides engineering suppliers there were suppliers of services, beginning with basic physics and chemistry research, economic analysis, financing, architecture and engineering design, transport, testing and inspection, radiological protection and safety, security, insurance, and of course marketing and public relations.

At the beginning of 1975 the nuclear industry's sales were settling down after rapid expansion, because of straitened circumstances in familiar markets, especially in the USA. But the competitive position of nuclear power as an energy supply technology looked better with every increase in the price of oil, with every pay-rise for coal miners, with every interruption of supplies of natural gas. The basic technology was well-understood, with a rapidly-growing body of experience ever more broadly disseminated. But all was not well with the industry. The electrical utilities, main customers of the nuclear industry, were themselves in acute financial difficulties, which rebounded onto the nuclear industry in the form of cancellations and delays to power station orders. But it is probably true that the industry's worries were more deep-seated, springing from a belated realization that public unease about nuclear power was not confined to pockets of out-and-out crankiness, but had become widespread and well-informed, among increasingly influential sections of society. The nuclear industry had been used to going its own way, carried by the momentum of its technology, paying only passing attention to criticism. But the criticism was becoming impossible to ignore. It was now undeniably influencing business decisions, to the detriment of the nuclear industry.

The Atomic Industrial Forum, the largest nuclear trade association in the world, prepared itself to mount a major campaign in 1975 on behalf of nuclear power. A group of twenty-eight leading US scientists led by the Nobel Prize-winning nuclear physicist Hans Bethe and by the long-time nuclear commentator and critic Ralph Lapp, plus nine other Nobel Laureates, published on 16 December 1974 a short, sober letter concluding: 'On any scale the benefits of a clean, inexpensive and inexhaustible fuel far outweigh the possible risks. We can see no reasonable alternative to an increased use of nuclear power to satisfy our energy needs.' But the nuclear opposition was able to point to Nobel Laureates who disagreed flatly with this proposition. It was clear that, in the months and years to come, the reactor debate would continue to reverberate. It was likewise clear that the fundamental considerations were not those of marginal costs, of one energy source against another. There were uncertainties and hazards associated with coal, with oil, with natural gas, with other sources of energy, with interruption of energy supply. The nuclear option had to be evaluated in these contexts – and in others, even less easy to quantify. It would not be a matter of balance sheets, but of human values: that is, of ethics.

8. Counting the Costs and Costing the Counts

Cost-efficiency was not a major criterion in the Manhattan project. What mattered was whether the technology could produce a bomb at all, and then whether it could produce a bomb in time; both of these criteria were of course satisfied. From then on the aim was to achieve, as one American general put it, 'a bigger bang for a buck'. But the early years of nuclear development, in the USA, the USSR, Britain and France, preoccupied almost entirely with military applications, were insulated to a large extent from the cold wind of everyday economics. Even so, those concerned with peaceful applications saw from the outset that basic economic criteria must be met.

One of the first coherent attempts to put numbers to civil nuclear proposals was that of R. V. Moore of Harwell in Britain, who in 1950 drafted a paper entitled 'Natural Uranium Reactors: Economic Factors in Power Production'. Moore correctly identified most of the quantifiable factors which would still be important a quarter of a century later. The accounting procedure Moore employed was the standard routine for costing a project like a power station, more or less like the one used today. It is particularly useful for the comparison of different investment options – for instance a nuclear power station and a coal-fired station of the same electrical output. In a moment we shall indicate how Moore and later analysts used the costing technique, and what results they got. First we describe the procedure.

The saleable product from a power station is electricity. The cost of producing it is partly the cost of building the station, and partly the cost of operating it. The cost of building the station – the 'capital' cost – is an investment, and entails interest charges year by year, as the capital value is gradually written down until the end of the station's useful life. Such capital 'carrying charges' can be shared out among the units of electricity generated, as part of the cost per unit. So can the operating costs – for fuel, labour, maintenance, insurance, taxes, et cetera.

In principle, then, it is easy to assess the cost of each unit of electricity (kilowatt-hour) produced by a station, and on this basis to decide whether the station is a good investment, on its own terms and by comparison with other alternatives. In practice, it is anything but straightforward. For a variety of reasons a station does not operate at its maximum rated design output all the time. Accurate costing must allow for its actual performance, and how well this performance approximates to the ideal embodied in the design.

A common measure of performance is a station's 'load factor' – the fraction of maximum possible operation it actually achieves during a certain period. This is sometimes stated as an 'availability factor', or as a 'capacity factor'. 'Availability' is the fraction of time that a power station is 'available' – that is, able to operate. It does not, however, identify periods when the station is operable but is for one reason or another supplying less than its rated maximum output. 'Availability' is not measured consistently. If a station is shut down for refuelling, the availability may be given as a fraction of the entire year. Or it may be given as a fraction only of that part of the year not occupied by refuelling; this gives a higher figure for its availability.

A more valuable measure of performance is the 'capacity factor': the total number of units sent out, as a fraction of the number which could be sent out if the station were to operate at maximum power for the entire year. The capacity factor takes account of low-power operation as well as of total shutdown.

Low-power operation of a station may be a consequence of operating problems, or of restrictions imposed by licensing authorities. It also results from the operational characteristics of the whole electrical system to which the station belongs. An electrical utility like Britain's Central Electricity Generating Board operates stations in what is called a 'merit order'. Those stations with lowest total generating costs operate all the time at maximum output, supplying the electricity which is needed all day long and all year long. Such stations are called 'base-load stations'. When the demand for electricity exceeds the base load – for example on cold days, and at meal-times – additional stations are added to the system to cope with the peaks; they are called 'peaking stations'. The order in which stations are added, which corresponds roughly to the order in which their running costs increase, is called the 'merit order'.

In general, stations with high capital cost but low operating costs are used as base load stations, for capital charges have to be paid whether or not the stations are in use. Nuclear stations, with their enormous capital cost, are at present invariably used as base load stations. Accordingly, nuclear stations are expected to operate at maximum output continuously. However, as electrical supply systems build more nuclear capacity, and as stations age, there will come a time when older nuclear stations drop lower in the merit order, and operate below maximum output at least some of the time. The resulting drop in capacity factor will make capital carrying charges of nuclear stations increasingly burdensome in later years; but the precise effect is difficult to anticipate.

The capital cost of a nuclear station has always been considerably more than that of a fossil-fuel station of equivalent output. But analysts from Moore onwards have shown that the running costs of the nuclear station will be less than that of the fossil-fuel station – sufficiently less, under the right conditions, to make nuclear electricity cheaper.

The fuel costs of a fossil-fuel station have usually been pretty unambiguous, whether the fuel be coal, oil or natural gas. Fossil-fuel costs must also take into account transport of the fuel to the station, and – in the case of coal – ash removal.

Nuclear fuel costs must be appraised differently. It is not merely a question of extracting uranium from the ground, transporting it to the power station and spraying it into the reactor. The uranium must be fabricated into fuel, whose design has a major influence on the amount of energy which can be generated from a given amount of uranium. We have already mentioned (p. 55) the desirability of high temperatures which give good steam quality and hence high thermal efficiency, and high fuel burn-up – longer-lived fuel elements. These factors affect the amount of saleable electricity produced from a given input of uranium. The fuel cost includes not only the cost of the uranium ore, but also the very considerable additional cost of incorporating it in a fuel element. If the reactor design requires enriched uranium, the cost of enrichment must be included, and is the largest single factor in fuel cost. A further additional cost arises when the spent fuel is reprocessed for recovery of usable fissile material and segregation of the radioactive residue.

Nonetheless, the extraordinarily concentrated energy in uranium compensates for the additional processing. Using the figures available in 1950, Moore found that electricity generated with the available nuclear technology could be comparable in cost to that from coal – about 0.5 old pence per unit. Moore did not assign any value to the plutonium which would be extractable from the used reactor fuel; but he suggested that in the long term only a system which made use of this plutonium would be of economic interest, because of the limited supply of uranium. Such considerations of course also influenced the early British assumption that only a fast breeder system, utilizing uranium-238 as well as the rarer fissile uranium-235, would be satisfactory.

Throughout the 1950s and much of the 1960s the economic status of nuclear power depended entirely on the current prices of alternative fuels – coal, oil and natural gas – in national economies. In the USA, as indicated in Chapter 4 (p. 129), the abundance of indigenous oil and gas, as well as coal, kept prices low; accordingly the 'break-even' price – at which nuclear and fossil-fuel electricity cost the same – was low enough to make the nuclear option economically doubtful. In Europe – particularly in Britain and France – such was not the case. Coal was available but not, it seemed, in enough quantity to meet the anticipated rapid upsurge in demand for electricity. Nonetheless it was apparent that any move to nuclear power would have to be gradual, and that its cost would not undercut that of coal for some years. Furthermore, it was also apparent that at least a few uneconomic stations would have to be built to give an adequate foundation for eventual economic nuclear power. Both Britain and France found a convenient compromise, building Calder Hall, Chapelcross, and Marcoule G-1, G-2 and G-3 for the military purpose of plutonium production and the civil purpose of electricity generation. In the USA, where uneconomic reactors seemed even more inevitable at current fossil-fuel prices, the AEC underwrote the construction of the entire first generation of US civil reactors – eight of which duly proved so uneconomic (or inoperable) that they were shut down permanently by 1970.

But in Britain the first generation of civil reactors proved, in due course, to have been an excellent investment. Despite high capital cost, the Magnox stations were so reliable in service that their running costs by the 1970s were comparable with the best of the fossil-fuel stations. Even after the discovery of unexpected corrosion effects, which required 'derating' the reactors by reducing the maximum operating temperature, the Magnox stations stayed among the best of the whole British electricity system. After 1970 when world oil prices began their stratospheric rise, the Magnox stations looked better than ever, except for the last and largest, Wylfa, which was beset with problems from the outset.

The cost of a power station of course depends on its size – the larger the more expensive. However, the cost does not increase in proportion; doubling capacity does not double the cost. The relevant measure is the cost per kilowatt of capacity; larger stations cost less per kilowatt – with certain provisos. Back in 1950 Moore's 90-MWe station was said to cost £9 million, not including the initial fuel charge: that is, £100 per kilowatt. Of the first generation of Magnox stations, the smaller early stations like Berkeley and Bradwell showed costs of around £180 per kilowatt; Wylfa dropped substantially lower, but not as had been hoped to less than £100 per kilowatt.

The second UK nuclear power programme was heralded in April 1964, in a White Paper which effectively conceded that the Magnox design had run its course, and that more compact enriched-uranium designs would now have to be considered. Among these were the British advanced gas-cooled reactor (AGR) and the American light water reactors (PWR and BWR).

Only a few months earlier, in December 1963, Jersey Central Power & Light had ordered the 640-MWe Oyster Creek boiling water reactor from General Electric, with no AEC subsidy; the time for economic nuclear power in the USA seemed at last to have arrived. However, the cost figures stated at the time proved in due course to involve a subsidy not from the AEC but from General Electric, who clearly regarded the sale as a 'loss leader', to induce further purchases and – presumably – to bring down their own costs to a point at which such sales would return a worthwhile profit. The successful Oyster Creek bid from General Electric was for a plant of 515-MWe output capacity, to cost $134 per kilowatt – with the added claim that the plant could in fact produce 640 MWe, reducing the cost per kilowatt to an impressively low $108. In the following three years, as indicated in Chapter 6 (p. 178), the hitherto reluctant utilities began queueing up to order light water reactors – more than thirty of them, with sizes escalating past the 1000-MWe mark, even including two 1065-MWe reactors for the Tennessee Valley Authority, in the heart of coal-mining territory. It looked as though the overall costs of nuclear stations had at last dropped below those of their competition.

In Britain the government announced on 25 May 1965 that the second nuclear programme would be based on the advanced gas-cooled reactor, and the CEGB in August ordered the Dungeness B station which was to prove such a disaster (see pp. 183–4). The CEGB published a report analysing the financial case for the Dungeness B AGR station, comparing it with the next closest bid (for a General Electric boiling water reactor); the Wylfa Magnox station; and the Cottam coal-fired station, situated in the region of lowest coal costs. In view of the fate which was to befall Dungeness B it is poignant to see the CEGB's cost calculations, to three significant figures. Taking account of load factor, life-time and interest rates, the CEGB was able to demonstrate to its own satisfaction that the Dungeness B station would supply electricity at a generating cost – capital plus running cost – of 0.457 old pence per unit, as against 0.489 old pence for a BWR of the same size. It gave no overall figures for the Magnox and coal-fired stations, but their effective average annual cost to the system, operating in merit order for plant lifetime, were stated to be for Wylfa £12.5 per kilowatt per year, for Cottam £9.0 per kilowatt per year, and for Dungeness B £7.8 per kilowatt per year. The capital cost of the stations were given as £124 per kilowatt for Wylfa, £43 per kilowatt for Cottam, and £92 per kilowatt for Dungeness B. All three stations subsequently failed to achieve their scheduled performance, with Wylfa accomplishing only an unimpressive 26 per cent load factor for 1973 and 24 per cent for 1974, Cottam beset by turbogenerator problems and Dungeness B seriously regarded by some as a potential total write-off; it is tempting to allude to the best-laid plans of mice and electrical utilities.

By December 1973 the CEGB, in support of its proposal for a vast programme of American light water reactors (see pp. 200–2) was claiming costs of quite a different order. The Parliamentary Select Committee on Science and Technology were told that, on the appropriate assumptions, Magnox electricity would cost 0.72 pence per unit, from a station costing £116 per kilowatt; advanced gas-cooled reactor electricity would cost 0.57 pence per unit from a station costing £89 per kilowatt; steam-generating heavy water reactor electricity would cost 0.51 pence per unit, from a station costing £67 per kilowatt; and pressurized water reactor electricity would cost 0.46 pence per unit, from a station costing £50 per kilowatt. These capital costs are for the nuclear steam supply system only, not for the whole station; the CEGB further complicated comparisons by quoting the total 'present worth' cost of the alternatives, an accounting procedure that reckons all outlays and incomes forward and backward in time with suitable interest or discounts, to bring them to a single date. Present worth calculations are however susceptible to the sort of fiasco already so grimly demonstrated by the second nuclear power programme, when the timing goes out of kilter, as well as to changing interest rates. The CEGB's exercise was in any case made somewhat academic by the subsequent government decision on a much smaller and later programme, whose financial status remains to be seen.

In the USA, the nuclear industry was watching intently the progress of generating costs, which were climbing on all fronts, fossil fuel and nuclear alike. The industry referred to a 'learning curve'; it was assumed that the initial availability and capacity factors of a new plant or size of plant would be low, but that as the learning process identified and extirpated the bugs, the performance figures would improve to an availability of better than 80 per cent and a capacity factor almost as high. In reality they however remained stubbornly much lower. In May 1974 the AEC published a report on nuclear power plant availability and capacity statistics for 1973. For the twenty-seven light water reactors which had been declared to be in commercial operation for at least three months during 1973, the availability factor was 70 per cent, and the capacity factor a mere 58 per cent.

In November 1974 David Comey, of Businessmen for the Public Interest, wondered, in the *Bulletin of the Atomic Scientists*, 'Will Idle Capacity Kill Nuclear Power?' His speculation was based on analysis of the reasons for the continuing low availability and capacity for US light water reactors. Comey concluded that the 'learning curve' did indeed bring availability and capacity up, to about 70 per cent by the third and fourth years of a plant's operation. But he pointed out that thereafter the performance showed a sharp drop, due not to recurring problems of youth but to problems of premature ageing: corrosion, fatigue, accumulation of radioactive crud complicating maintenance, and the like. Comey's analysis showed that plants reaching the seventh year of operation were clocking in with average annual capacity factors well below 40 per cent – clearly an alarming development for the industry, if it were to continue.

In early 1975 the US Nuclear Regulatory Commission

(successor to the AEC) announced that forty-two commercial nuclear plants operating in 1974 had an average availability of 68.5 per cent, and an average capacity factor of 57.2 per cent – by any criteria nothing to write home about. By basic economic criteria nuclear power looked to be in difficulty.

By the late 1960s, moreover, it had become apparent that the costs of electricity production did not all show up in the accounts. Extracting coal killed and maimed miners, and ravaged the landscape; burning it spewed sulphur dioxide by the thousands of tonnes into the air, producing health hazards in every breath. Extraction and transportation of oil produced disasters like the *Torrey Canyon* and the Santa Barbara platform, coating coastlines with black filth and wreaking long-term havoc on marine ecology. Natural gas was less fraught with obvious unpleasant side-effects, but was unfortunately likely to be exhausted within a generation, and – like coal and oil – might better be used as raw material for chemical processes, for which the fossil hydrocarbons were (and are) irreplaceable. For a brief time such arguments reinforced the move towards nuclear power, which seemed by comparison clean, safe, and available into the indefinite future. But the nuclear honeymoon was fleeting.

Certain aspects of nuclear power, of profound economic significance, had also been left out of the reckoning. There was, to begin with, the vast background of research and development, whose cost had been borne almost exclusively within military budgets everywhere. There were the all-important fuel cycle facilities: uranium mines and mills, hexafluoride conversion plants, enrichment plants, fuel fabrication plants, fuel reprocessing plants, waste storage installations and transport systems – all developed and built with military funds, for weapons production. As the civil applications of nuclear energy grew more and more prominent, efforts were made to draw a clear distinction between the civil and the military sides of the undertaking; but the results were unconvincing. Only in the 1970s, as the installations built for military purposes were reaching the end of their useful lives, did it become possible to anticipate cost-accounting which would incorporate suitable allowances for the cost of capital installations elsewhere in the fuel cycle.

Unfortunately for the industry, by the end of 1974 the civil fuel cycle situation was looking precarious, especially in the USA. As we have already indicated, all three of the civil reprocessing plants were unavailable until 1977 at the earliest and the Midwest Fuel Recovery Plant of General Electric looked to be a $65 million write-off. Non-governmental enrichment facilities were having serious difficulties getting beyond the talking stage. Although given access to classified military data on enrichment technology, private industry was unwilling to contemplate an eight-year construction period for an enrichment plant, with no return on investment, and with interest rates in double figures. Influential nuclear spokesmen in the USA, led by long-time Congressional Joint Committee member Craig Hosmer, were urging the setting up of a US Enrichment Corporation – a nationalized enrichment facility. In almost any other context a nationalized energy facility in the civil sector in the USA would be unspeakable heresy; but it may yet materialize.

Elsewhere the notion of nationalized energy facilities is of course less unthinkable. British Nuclear Fuels Ltd, the fuel cycle service at present wholly owned by the United Kingdom Atomic Energy Authority, announced in late 1974 plans to invest some £500 million in new enrichment and fuel-reprocessing capacity. (The British government in late 1974 refused a request for financial assistance to help British Nuclear Fuels build a pilot plant for solidification of high-level waste; funds will thus have to be found within the company's own accounts if this technology is to be introduced.) For several years British Nuclear Fuels, in partnership with West German and Dutch firms, has been developing gas centrifuge enrichment technology. The joint firm of CENTEC has been building centrifuges, and URENCO has begun constructing centrifuge enrichment plants at Capenhurst in Britain and at Almelo in the Netherlands. France, meanwhile, has undertaken construction of a new civil gaseous diffusion plant at Tricastin, with other European partners. The Soviet Union has also begun to offer toll enrichment services to foreign buyers. It is not past conception that the USA may in the 1980s be importing enriched uranium, after its long-time role as the exporter. But it is safe to assume that – unlike the early days, when the AEC supplied enriched uranium for civil purposes at rock-bottom prices – future suppliers will charge what the market will bear, whoever the buyer may be. A foretaste of this came in 1973–4, when the AEC demanded that overseas buyers make firm contracts eight years in advance, with severe penalty clauses – a move interpreted as an attempt to head off a major switch to European suppliers of enriched uranium. Prices of separative work have soared, from less than $30 per kilogram separative work to a price of $100 per kilogram quoted by British Nuclear Fuels in early 1975 for delivery in the early 1980s.

British Nuclear Fuels and other enrichers are now beginning to require prospective buyers to commit themselves early enough to underwrite the capital investment their order requires. If such a procedure becomes common, nuclear operating costs are bound to rise; it will also link the entire industry inextricably together, so that a miscue anywhere will reverberate through the whole international establishment.

Supplies of uranium itself have begun to cause concern in some quarters. The long-standing base price of $6 per pound of yellow-cake (U_3O_8) – that is, some $13 per kilogram – has climbed past the $10 mark and is on its way to $20 for contracts recently negotiated, especially those involving supply over a period of years. Projections of nuclear capacity in the USA, in Europe and world-wide, juxtaposed against known reserves of uranium of varying grades and costs, have been declared by some commentators to indicate a probable shortfall of supply before the turn of the century, unless a major move is made into plutonium-fuelled reactors. A suitable mix of breeders and thermal reactors would be able to utilize almost all the uranium, instead of only 1 or 2 per cent of it, but it would involve extra costs in extracting and refabricating fissile material repeatedly.

But other commentators point out that the cost of uranium ore is only a very small part of the cost of electricity produced by a nuclear power station, and that accordingly even a rise to over $50 per pound for yellow-cake would produce only a small increase in the cost of electricity generation. In 1972 reasonably assured world resources at a price of less than $10 per pound of yellow-cake amounted to some 866 000 tonnes of uranium. A major OECD report on 'Uranium: Resources, Production and Demand', published August 1973, giving these figures, described the upsurge in exploration activities in many parts of the world, and analysed possible production and consumption rates, given projected increments of nuclear generating capacity. The report suggested that annual production of uranium would be of the order of 50 000 tonnes per year by 1978, with demand reaching 60 000 tonnes per year by 1980 and 120 000 tonnes per year by 1985.

Such extrapolations are of course acutely sensitive to the timing, as well as the scale, of increase of nuclear capacity, and to the possible introduction of plutonium recycling in thermal reactors. Recent plant delays, especially in the USA, and fall-off in growth of electricity demand, doubtless require a re-examination of the position. But future uranium supply will clearly have to be watched, especially since reactors ordered in 1975 will still be presumed to require uranium until at least 2005 AD.

Perhaps the most controversial financial aspect of civil nuclear power is insurance against third-party liability in the event of accident. In the mid 1950s, as we have described, prospective operators of nuclear power stations in the USA shied away, worried at the monumental sums they might have to pay out in the aftermath of a major reactor accident. The numbers put forward by WASH-740 (see pp. 170–2) were well outside actuarial experience. On the one hand, the probability of an accident was certainly declared to be very low. But on the other, the awesome consequences foreseen, should this improbable occurrence nonetheless occur, involved sums of money between ten and one hundred times the largest previously contemplated for third-party liability. Without even referring to the human dimensions of such a disaster it was clear that the largest utility in the country would be wiped out financially long before covering its liabilities – and so would most insurers.

The deadlock was broken by the passage of the Price–Anderson Act of 1957, which established the principle of an upper limit to operator liability in claims for third-party damages, and provided a Federal umbrella of extra coverage (see p. 172) – initially $60 million from private insurers, plus $500 million from the government. The Price–Anderson Act also prompted some rather specialized provisions in the insurance policies subsequently drawn up between reactor operators and the pools of insurers who came together to underwrite nuclear risks. To mention only one: after a set period, usually ten years, provided an operator has not claimed against his policy, his premium payments are refunded – an arrangement which would delight any householder, but which is only available if he has a reactor on the premises.

In Britain the Nuclear Installations (Licensing and Insurance) Act 1959 established a like limitation on operator liability – at the even lower figure of £5 million from private insurers, plus £43 million – raised by a later Act to £45 million – from the government. Similar provisions apply elsewhere. Since the consequences of a nuclear accident could not be expected to observe national boundaries, attempts began in the late 1950s to codify an international convention on third-party nuclear liability. In 1975 the attempts continue. Since other industries, including other energy-supply industries, must provide their own third-party coverage out of working funds, there is little doubt that the nuclear industry gains a distinct competitive advantage from the present provisions.

There is, however, an intriguing situation arising in the USA. The Price–Anderson Act was renewed in 1965, two years ahead of time, to avoid any last-minute awkwardness; and a similar move was made in 1974, before the most enthusiastic nuclear proponents, notably Congressmen Craig Hosmer and

Chet Holifield, retired from Congress and thus from the Congressional Joint Committee. The 1974 move unexpectedly backfired: the newly installed President, Gerald Ford, took exception to a rider requiring Congressional approval of the as-then unpublished WASH-1400 Reactor Safety Study, and vetoed the Price–Anderson renewal. Congress broke up without further action; and the incoming Congress proved to be much more sceptical about the nuclear option than its precursors. As this is written plans are under way for new Congressional hearings on the Price–Anderson provisions and other aspects of governmental backing for nuclear energy. It is possible that the Congress will take WASH-1400 as a basis for suggesting that, since reactor accidents seem to be so remote, and since the consequences seem not to be quite so terrible as previously thought, perhaps now US reactor operators will be prepared henceforth to finance their own insurance coverage *in toto*. It will be a most interesting colloquy.

Support and subsidy for research and development, fuel cycle services and insurance are of course financial matters, and inherently quantifiable if not easy to quantify. Other factors also fail to show up on a nuclear balance sheet, factors which seem inherently impossible to quantify. There is, to be sure, a stratum of numerical argument about managing high-level radioactive wastes. But the true costs of producing such wastes and bequeathing them to countless future generations are not monetary but ethical. In a similar, if more immediate, fashion, the costs of designing an economy which creates and transports potential nuclear weapons materials like plutonium, by the tonne, with the concomitant security risks, are not monetary but social. The costs of radiological protection for nuclear workers are highly visible, capital and running costs which show up in corporate accounts. But many people now consider that the invisible costs associated with nuclear energy may be the hardest to bear.

Curiously enough, this generalized uneasiness may, after all, have an identifiable financial manifestation. In late 1974 Irvin Bupp of Harvard Business School and Jean-Claude Derian of the Center for Policy Alternatives at MIT drafted an analysis of *Trends in Light Water Reactor Capital Costs in the US: Causes and Consequences*. Their analysis led them to a startling conclusion. The cost per kilowatt of nuclear stations was shown to be increasing, not decreasing – although a decrease was to be expected if the new larger stations were displaying the anticipated economies of scale. On the data available Bupp and Derian pointed out, for instance, that reactors ordered in 1968, expected to cost only $180 per kilowatt, were in fact costing about $430 – well over twice as much. Furthermore, the difference between costs anticipated and costs eventually realized was continuing to widen. Estimates in 1973 suggested that plants entering service in 1982–3 might cost some $700 per kilowatt; but Bupp and Derian insisted that such estimates were impossible to deduce with confidence from the data, and were in fact 'little more than an educated guess'. What was unambiguously clear was that capital costs of large light water reactors showed no signs of stabilizing, and indeed were still climbing at alarming rates, so much so that the lower fuel-cost of nuclear power stations was in danger of being offset by high capital costs.

More unexpected still were the reasons Bupp and Derian advanced for the cost escalation. They found no correlation between increased cost and other likely factors, such as construction delays, or length of construction period. But they found a strong correlation between the time taken to license a reactor and the eventual capital cost per kilowatt. For reactors taking the same total time from licence application to criticality, the longer the part of the time occupied by licence proceedings the higher the eventual capital cost per kilowatt. Bupp and Derian were led to deduce that public opposition, as exerted during the licensing process, was adding a 'social cost' factor. This additional cost was interpreted by Bupp and Derian as 'forcing a downward revaluation of the social value of reactor technology'.

If, in some fashion, the 'hidden' costs of nuclear technology become part of the visible accounts, energy planning will be much better able to evaluate the various options available, in numerical terms. But to many the ethical issues will remain unambiguously ethical issues, whose costs as such are simply irrelevant.

9. Plutonium at Large

Plutonium is man-made, an element which did not exist in nature until 1940. Glenn Seaborg and his colleagues at the University of California, using a particle accelerator, first created plutonium, effectively one atom at a time. Seaborg later recalled keeping the world's entire stock of plutonium in a matchbox in his desk. The physical and chemical properties of plutonium were analysed using quantities invisible to the naked eye. Left to itself, a small amount of plutonium undergoes gradual alpha decay, with a half-life of 24 400 years for the commonest isotope, plutonium-239. Curiously enough, when plutonium-239 emits an alpha particle it becomes uranium-235, whose most spectacular property is also exhibited by plutonium-239: the ability to sustain a chain reaction.

Within months of Seaborg's original work it became apparent to the insiders that plutonium-239, like uranium-235, would be potential raw material for a nuclear bomb; and would in some respects be even better than uranium-235. As we have already described, the production of plutonium in quantity was a principle aim of the Manhattan project and of post-war efforts in the USA, the USSR, Britain and subsequently France. However, as well as the fissile property which made it militarily desirable, and the most concentrated source of energy available, plutonium fairly soon showed other attributes of singular unpleasantness. Like radium it proved to be a ferocious radio-active poison, dangerous in quantities of a microgram or less. Plutonium produced in a reactor was – unlike uranium – mainly fissile nuclei, and could without warning reach criticality, either in solid form or in solution. Plutonium workers noted the eerie phenomenon of plutonium oxide 'breathing'. If a tray of the fine fissile dust was filled to criticality, its surface pulsed gently, expanding with the energy release at criticality, thereby going subcritical, collapsing back to criticality, and repeating the cycle endlessly – and of course emitting a burst of neutrons with each lapse into criticality.

What made – and makes – plutonium a matter for urgent concern is not merely its radioactivity, or its fissile nature; other 'actinides' are both more radioactive and more readily fissile. But plutonium now exists in quantities measured in tonnes, is being created daily in ever-larger amounts, and is expected to be a major constituent of the nuclear fuel cycle of the future, not merely as a product but as a raw material. The two hazards, the radioactive and the fissile, are becoming increasingly serious. The past history of plutonium management cannot be called reassuring.

The Toxic Hazard

At an early stage in the short chronicle of plutonium its anti-social nature was fully appreciated; nuclear scientists always treated it with profound respect. But in processing on an industrial scale it is difficult to keep track of every last microgram, even when a microgram may be lethal. Technicians working with plutonium inevitably grew to take it for granted – not precisely blasé about it, but less than exacting about the precautions taken. In Britain, by the late 1940s, more than one scientist had been heard to declare that, should he happen to get plutonium on a cut finger, he would at once cut the finger off. But by the 1960s Edward Gleason didn't even know he'd met plutonium – although it might have been as well if he had taken similar drastic measures.

There has not been much 'convincing' evidence of plutonium toxicity in human beings; because of the long delay before the resulting effects become clear, only the most massive overdose – which in the case of plutonium may be in micrograms – shows up soon enough to be clearly attributable to plutonium. But Gleason's case seems grimly convincing to many.

The plutonium in question was a contaminant in a solution of gold chloride, owned by the AEC, being shipped in a forty-litre glass carboy, as Lot 61, from the Nuclear Materials and Equipment Corporation to the Brookhaven National Laboratory, between 4 and 14 January 1963. The official AEC report, dated 11 April 1963, by Anson Bartlett of the Division of Inspection, included lengthy interviews with the series of people whose paths crossed that of Lot 61. It is a thirty-seven-page chronicle of casual carelessness. Edward Gleason just happened to be the unlucky one.

Wrongly labelled, wrongly packaged and transported in contravention of regulations, it came to Gleason's attention at the Eazor Express Corporation trucking terminal in Jersey City on 8 January 1963. Gleason noticed that the crate containing Lot 61 had been leaking; as he manhandled it off the trailer, liquid dripped from the crate onto the trailer floor. When a puddle formed he turned the crate over on to another side and the leaking stopped. Gleason did not know, as he testified later, that the box contained radioactive material, much less that it contained a solution of plutonium in a glass carboy with a loose stopper.

Cleaning up the long trail left by Lot 61 took the AEC some time, and many thousands of dollars. Unfortunately they were unable to clean up the droplet that apparently seeped through a tiny cut on Gleason's hand. By 1966 it was evident that Gleason had developed cancer: a soft-tissue sarcoma, which by 1968 necessitated amputation of his left hand and arm and part of the shoulder. Subsequent radiation treatments failed to halt the spread of the sarcoma. Gleason sued the AEC and the companies involved in the shipment of Lot 61 for two million dollars in damages. The AEC and the companies contested the suit on the basis that there was no conclusive evidence that Gleason's sarcoma had resulted from his encounter with Lot 61. US vital statistics indicate that death from such a cancer is likely to occur fewer than once in a million persons per year. After a lower court decision against him Gleason died; his widow has carried on the legal challenge, against the defendants' stubborn insistence that Gleason's cancer had been a one-in-a-million bit of bad luck.

Fortunately there seems thus far little evidence of occupational hazard to plutonium workers in the nuclear industry, military or civil. This may be because the industry has always taken stringent precautions; in part, at least, it may be because little serious effort has been made to keep track of people who have worked with plutonium. Since the effects of exposure to plutonium may not be as dramatic as those suffered by Gleason, and may take many years to become manifest, their cause may not be recognized. In 1968 the AEC set up at Hanford a Transuranium Registry, an attempt to follow up the medical life-histories of employees who had worked with plutonium and other alpha-emitting substances. The difficulties, including that of intrusion on personal privacy, must not be minimized; but without some such effort, not only in the USA but in other countries with major nuclear enterprises, no adequate data will ever be assembled on the human pathology of plutonium – not, at least, until it is too late to do anything about it.

It is already too late to do anything about the plutonium added to the global environment by past weapons tests and by mishaps, of which there have been several.

In August 1964 a US space vehicle powered by a SNAP 9A radioisotope generator, which used the heat generated by radioactivity to power a thermoelectric system, plunged back into the atmosphere and burned up. The disintegration of the generator added 17 kilocuries of plutonium-238 to the atmospheric burden of radioactivity from weapons tests. (Plutonium-238, with a half-life of 87 years, is a more intensely radioactive sibling of plutonium-239, with a much higher specific activity.)

In January 1966, to the chagrin of the USA and the consternation of Spain, a US B-52 bomber carrying four hydrogen bombs collided with a tanker aircraft and crashed near the Spanish village of Palomares. There was no thermonuclear explosion; but high-explosive in the trigger mechanism of one of the bombs blew up and threw plutonium in all directions. The USA at once dispatched a clean-up battalion of decontamination experts, radiological safety investigators, and a cast of thousands, complete with all manner of hardware. Some 2.4 square kilometres were stripped of crops and ploughed to a depth of 25 centimetres to bury any finely-scattered plutonium. From the patch of worst contamination, some 2.2 hectares, all the vegetation and soil was removed and packed into drums, for removal to the AEC facilities at Savannah River for burial. One of the other hydrogen bombs had to be dredged by midget submarine from the bed of the Mediterranean off the Spanish coast.

In January 1968, a re-run of the 1966 Palomares incident occurred, when a second B-52 crashed, this time near Thule airbase in Greenland, presenting the US authorities with another 'problem of large area plutonium contamination', in the words of a leading US expert on the subject.

Back home in the USA plutonium continued to cause problems – particularly at the AEC's plutonium plant at Rocky Flats, about thirteen kilometres outside Denver, Colorado. Rocky Flats, then operated by Dow Chemical for the AEC, for the fabrication of fissile components of nuclear weapons, had a history of leaks, spills, fires and explosions. In May 1969 it set a record, with the most expensive industrial fire in the history of the USA, possibly of the world. The fire in question was just a fire, not a conflagration (of which there have been examples like the razing of Chicago, whose total cost might have been greater). In fact from the outside it was impossible to tell that a fire had even occurred. But it had.

Since, as we have described, plutonium is toxic in microgram quantities, pyrophoric and susceptible to criticality, it might be expected that it would be handled with unparalleled delicacy. Such does not, at Rocky Flats, appear to have been the case.

Work on plutonium is carried out inside 'glove boxes': sealed boxes with windows, with portholes into which are clamped heavy rubber gloves. The plutonium stays inside the line of glove boxes while workers outside thrust their arms into the gloves to carry out operations like forming, pressing, milling, machining, polishing or calibrating the plutonium metal shapes required for bombs. After a number of lesser fires, a glove-box

fire on 11 September 1957 caused damage costing $818 600.
Two fires in 1965 cost another $40 000. They also exposed
workers to airborne particles of plutonium oxide: the fire on 15
October 1965 gave twenty-five workers up to seventeen times
the maximum permissible exposure. Indeed exposure and
contamination incidents seemed to occur almost on a routine
basis, leading to a fierce confrontation within the union involved.
The epic of Rocky Flats is recounted in detail by Roger Rapoport (see Bibliography, p. 292); after a succession of fire and
contamination incidents it came to a climax on 11 May 1969.

Some scrap plutonium had been stored improperly in uncovered
cans under a glovebox in building 776-777 at Rocky Flats. That
morning, a Sunday, it did what plutonium has a nasty habit of
doing: it ignited spontaneously, and set fire to the glovebox itself.
'Glovebox' here gives a misleading impression of size; it was no
mere box but a large chamber, constructed of nearly six hundred
tonnes of combustible material, radiation shielding included. The
heat detectors installed to give an alarm in such an eventuality
had – with stultifying lack of foresight – been positioned outside and
under the glovebox, so that they were themselves insulated from
the heat until the fire was blazing out of control. When the detectors
did finally sound the alarm at 2.27 p.m. the smoke was so thick
that the arriving firefighters could barely find their way to the fire.
The smoke, of course, included an impressive concentration of
plutonium oxide; among the materials consumed in the blaze was
some $20 million-worth of plutonium. At $10 000 per kg that
works out to about 2000 kg – two tonnes of a material of which a
microgram is likely to be toxic. Needless to say the firefighters wore
breathing apparatus. Fortunately the fire did not burn through the
roof of the buildings. If it had, the consequences for the surrounding
area, including the city of Denver, would have been a 'problem of
large area plutonium contamination' beside which the Palomares
and Thule episodes would have been laughable. For the firefighters
the immediate problem was the possibility of criticality, if they
used water on the fire. Because of its effectiveness as a moderator
water is stringently contra-indicated for such occasions. But the

firefighters used up their carbon dioxide within ten minutes, and had nonetheless to resort to water. It took them four hours to bring the fire under control, and some trouble-spots burned throughout the night. When eventually, some days later, investigators with breathing apparatus could survey the radioactive ruins, they assessed the damage at $45 million – plus the $20-million-worth of plutonium. On 20 May a deputation from the AEC requested and received the necessary government funds not only to repair the damage but also to carry on with a major expansion programme.

The aftermath of the Rocky Flats fire stirred scientists of the Colorado Committee for Environmental Information, led by Dr Edward Martell, to investigate just how much plutonium had escaped from Rocky Flats. What they found was acutely disturbing. It was clear that plutonium had indeed reached some neighbourhoods in Denver; it was furthermore clear that some of the plutonium in soils near the plant had been there since before May 1969.

In February 1970 the scientists, under the banner of the Colorado Committee for Environmental Information, published their report; the AEC, while supporting its own studies to counter the outside evidence, met with the Committee scientists on 10 February and – according to Peter Metzger, who was present (see Bibliography, p. 292) – threatened to make trouble for the scientists with their Federal employers. However, after an April meeting – vividly recounted by Roger Rapoport (see Bibliography, p. 292) – between senior Congressional Joint Committee members, AEC staff and union representatives the AEC revealed on 21 July 1970 that waste contaminated with plutonium had indeed been buried both inside and outside the gates of the Rocky Flats plant – some 1405 barrels of it, which they had from 14 April onwards dug up and removed. Other areas of soil contaminated by plutonium were to be removed to an authorized burial site. Only one little difficulty remained: how to gather the contaminated earth without letting its radioactive constituents be picked up by the wind and blown into Denver.

In early 1972 the AEC confessed to another somewhat disconcerting discovery. At the Hanford reservation in Washington state it had long been standard practice to dispose of liquid wastes slightly contaminated with radioactivity by pouring them into trenches. The trenches had concrete walls but no floor. Liquid waste decanted into a trench soaked into the soil, carrying the smattering of radioisotopes down with it into what had been assumed would be effectively permanent subsurface burial. The soil particles would absorb the radioactive nuclei and hold them indefinitely, until their activity had decayed to insignificance. Among the radioisotopes thus poured into the soil was, over a period of years, some 300 kilograms of plutonium; of this about 100 kilograms was poured into trench Z-9. It was expected that the plutonium, like the other contaminants, would be dispersed through the soil in an insoluble form. It would then remain well above the water table in a sufficiently dilute form to cause no trouble, despite its 24 400-year half-life.

Unfortunately the procedure failed to take account of one factor. The soil beneath trench Z-9 did indeed absorb plutonium as anticipated; but it did so selectively. Like the standard chemical separation technique of column chromatography, it separated radioisotopes into layers of different species at different depths. One layer not far below the surface was found to be dismayingly rich in plutonium: so rich that heavy rain soaking into the ground might, with its moderating effect, trigger a nuclear chain reaction. The resulting energy release might throw up a 'mud volcano', spewing plutonium into the Hanford air. Investigators from the National Academy of Sciences pointed out that changes in ground-water chemistry might mobilize some of the buried plutonium, creating even heavier concentrations and a yet more serious threat of chain reaction. Accordingly, in the spring of 1972 the AEC asked Congress for – and got – $1.9 million to dig up trench Z-9: the world's first fully automated plutonium mine. Meanwhile, since more sophisticated disposal was not yet available, the AEC continued to pour dilute plutonium waste solutions into trench Z-18 near by.

T–K

The Fissile Hazard

Military facilities, of course, produce plutonium for weapons; civil nuclear systems also produce plutonium in quantity, with as yet no outlet for it. Some governments have operated a system whereby they 'buy back' all plutonium created in commercial reactors, crediting the reactor owner with its value against the charges for other fuel-cycle services such as enrichment. The US government provided a 'buy-back' service until 1970. Since that time the electrical utilities operating power reactors in the USA have been stockpiling the plutonium at storage facilities or 'depositories', awaiting authorization to incorporate the plutonium in fresh reactor fuel in place of uranium-235. Until that time they cannot legitimately credit their accounts with the plutonium value, which is of the order of $10 000 per kilogram. Since a 1000-MWe reactor produces well over 100 kilograms of plutonium per year, the value of the plutonium is something over $1 million per reactor per year. In August 1974 the AEC at last published its *Generic Environmental Statement on Mixed Oxide Fuel* – which concluded that plutonium could be used in light water reactor fuel without any particular unease. But the commercial traffic in plutonium associated with 'plutonium recycle' seems bound to exacerbate the already intractable problem of restricting access to fissile material.

Such restriction, policing the world's breeding grounds for nuclear weapons, was conceived as a main responsibility of the International Atomic Energy Agency, from its inception in 1956. But its success was then and remained limited. In March 1962 the Agency safeguards system came into being. An incidental difficulty was that it could only be exercised when a national government permitted it to be – and very few national governments were so inclined. It is hard to feel any surge of confidence about the essential efficacy of the present safeguards arrangements; and in fairness it must be said that this unease is shared by many Agency safeguards staff members.

During the first agreed moratorium on nuclear weapons testing, the Eighteen Nation Disarmament Conference had been set up, by a UN General Assembly resolution of 20 December 1961. Little discernible progress was made at the Conference towards disarmament of those countries already in possession of nuclear weapons and delivery systems. But it would clearly be preferable if, at the very least, countries without nuclear weapons could agree not to acquire them. By 1965 the weapons-possessors were putting forward proposals to head off 'proliferation' of nuclear weapons. US and Soviet draft treaties on 'non-proliferation' were presented to the Eighteen Nation Disarmament Conference – by this time usually just called the Geneva conference, and already showing every sign of becoming a permanent institution – and to the UN General Assembly. Reactions among the non-possessors ranged all the way from whole-hearted support to scornful rejection.

In the preceding decade a number of countries had found themselves confronting the decision: to have or not to have. In almost every case the internal domestic debate was heated. Sweden, whose first power reactor at Ågesta had gone critical in July 1963, gave serious thought to building tactical nuclear weapons with accumulating Swedish plutonium. Military leaders and conservative politicians were vigorous advocates of this policy; but the Social Democratic government, after several years of discussion, at last decided against the idea.

Israel determined in 1957, after the Suez débâcle with Britain and France, to secure the option of Israeli nuclear weapons. With French help – and aided by the continuing influx of highly qualified immigrants – Israel built a 26-MWt research reactor at Dimona in the Negev desert. The installation was always identified as being for research, but its operations were kept under a veil of military secrecy; and its annual plutonium output of five to seven kilograms could be equated to a bomb a year. In Israel the domestic debate about nuclear weapons was particularly bitter; the apparently perpetual jeopardy of the country among its Arab neighbours made the issue far from theoretical. The question was quite straightforward: would acquiring nuclear weapons make it easier or less easy to guarantee the future of Israel? No convincing answer was forthcoming either way. But Israel was in no mood to forgo the nuclear option in exchange for promises. Promises could be broken.

From the 1950s onward India developed advanced nuclear technology worthy of comparison with any. India's first research reactor at Trombay went critical in 1956; apart from the nuclear weapons nations only Canada, Norway and Belgium were in the field sooner. In 1960 the 40-MWt CIRUS reactor at Trombay, a joint enterprise between India and Canada, went critical. Canada worked in close partnership with India in nuclear matters throughout the 1950s and 1960s, although the heavy water for the CIRUS reactor, and also the first Indian nuclear power station at Tarapur, were supplied by the USA. India also acquired a plutonium separation plant. Over a period of years, India emphasized her interest in the use of nuclear explosions for civil engineering purposes. As a 'non-aligned' country India endeavoured to maintain her diplomatic distance from both nuclear weapons camps. When China, one of India's own most persistent foes, weighed in with a bomb, India made clear her intention of reserving the nuclear option, however vigorous the pious Indian denunciations of nuclear testing and brow-beating by the original nuclear weapons nations.

On 17 June 1967, China exploded its first thermonuclear weapon, of 3-megaton yield, at its Lop Nor test site in Sinkiang. It had taken China only two-and-one-half years to proceed from its first fission explosion to its first hydrogen bomb; the 17 June 1967 test was a true bomb, dropped from an aircraft.

On 12 June 1968, the UN General Assembly commended the joint US–Soviet draft Non-Proliferation Treaty. 'Commending', of course, was only an acknowledgement that such a document existed; it needed to be signed and ratified by interested national governments before any of its provisions could be regarded as having more than philosophical import. Article I of the Treaty prohibits the transfer of nuclear weapons (or other nuclear explosive devices, which might as well be weapons) to any states under any circumstances. Article II prohibits Treaty-members from manufacturing or acquiring nuclear weapons – but not from preparations up to the point where it is only necessary to fit a weapon together. Article III obliges non-possessors of weapons to accept International Atomic Energy Agency safeguards on all their nuclear activities, to ensure that they do not covertly 'divert' fissile material to make nuclear explosives; no Treaty member may supply fissile material to a non-member unless the non-member agrees to the International Atomic Energy Agency safeguards. Article IV says that all Treaty members may, nonetheless, do anything else they wish to do with nuclear energy for peaceful purposes, and may help each other to this end. Article V says that possessors of weapons must agree to provide nuclear explosives for peaceful purposes, under international control and for appropriate charges, to non-possessors who want them. Article VI exhorts members to keep on trying to dispense with nuclear weapons – to find 'effective measures for nuclear disarmament'. Article VII says that members may agree on 'nuclear-free' zones. Article VIII says that a conference to review the Treaty shall be held five years after it comes into force; the first review Conference was held in Geneva in May 1975. Article IX permits additional nations to become parties to the Treaty once it comes into force. Article X permits a member to withdraw from Treaty obligations on three months' notice if the member decides that extraordinary events, related to the subject matter of the Treaty, have jeopardized the supreme interests of the member-country – or, in less fancy language, you may withdraw in three months if you want to.

Taking Article X together with Article II, and noting that final assembly of a nuclear weapon need take much less than three months, if you have the parts ready and are in a hurry, the strictures of the Treaty cannot be called onerous. Yet only about half the nations of the earth have ratified it. Non-signatories include of course France and China, Argentina, Brazil, Chile, Cuba, India, Israel, North Korea, Pakistan, Portugal, Saudi Arabia, South Africa, Spain, Tanzania, North Vietnam and

Zambia. Other countries have signed, but then failed to ratify: among them are Australia, Egypt, Indonesia, Japan, South Korea, Switzerland, Turkey and Venezuela. Nations which have both signed and ratified the Treaty – thus denying themselves nuclear weapons – include Chad, Haiti, Lesotho, Madagascar, Malta, Nepal, San Marino, Swaziland and Togo.

On 18 May 1974, in the Rajasthan desert in the west of the country, India detonated a 15-kiloton underground nuclear explosion. India thereby became the sixth country to possess nuclear weapons technology, although Indian spokesmen insisted stoutly that their explosion was for 'peaceful purposes' only. Be that as it may, the Indian explosion underlined with dramatic abruptness the question which had begun to preoccupy many nuclear observers. In the burgeoning world-wide enthusiasm for nuclear energy, it had for some years seemed possible to draw a degree of distinction between civil nuclear systems and their military implications. As of 18 May 1974 such a distinction became very difficult to discern.

During the 1940s and 1950s there had been an aura of awesome mystery about nuclear weapons, about the technology, about its impact on policy, about the unthinkable consequences should such weapons be used. But somehow, by the late 1960s, the growing nuclear arsenals were taken for granted, no longer a matter for conspicuous public concern. Millions of people had been or were still employed in constructing and tending nuclear and thermonuclear bombs – and many thousands of them shared the detailed knowledge which only a decade before had been the most closely cherished secrets. What had been a vocation had become a job like any other. As the global inventory of fissile materials soared, the quality of their supervision dwindled.

The Nuclear Materials and Equipment Corporation reported to the AEC in 1965 that over some six years' activity at its fuel-fabrication plant in Apollo, Pennsylvania, it had somehow mislaid more than 60 kilograms of highly enriched uranium – enough weapons-grade material to make several fission bombs. This material unaccounted for – MUF, in industry jargon – might have been simply the cumulative amount of

traces not recovered from scrap. On the other hand, it might
not. The AEC forthwith set up a new Office of Safeguards and
Materials Management, charged with tightening up controls
on fissile materials.

When the first Chinese fission bomb went off in 1964, and
proved to be made not of plutonium but of enriched uranium,
dumbfounded Western nuclear observers at first assumed it
had been stolen – perhaps from Apollo? Satellite photographs
of the Chinese gaseous diffusion plant subsequently obviated
the need to assume Chinese theft of uranium-235. But the
uranium was still missing, although some was later found in
scrap; some of it has never been recovered. If it could be
imagined that a national government might steal fissile material,
what other grim possibilities might exist? On 27 October 1970
the police in Orlando, Florida, received an anonymous message
informing them that the sender had a hydrogen bomb, and
warning that it would be set off unless one million dollars were
paid. The following day another message included a sketch
diagram of the bomb – and distraught officials confirmed that it
looked all too genuine. There did not seem to be any theft of the
necessary nuclear material recorded; but it was also impossible
to be sure that no such theft had occurred. Then, to the bottom-
less relief of the local authorities, a police trap snared the letter-
writer: who turned out to be a fourteen-year-old boy, and whose
'bomb' was a hoax. But it might not have been.

In the summer of 1971 Kansas State University played host to
a deadly serious conference whose theme was 'Preventing
Nuclear Theft'; it was apparent that doing so would not be easy.
In September 1972 the annual international Pugwash conference of leading scientists from many countries, meeting in
Oxford, England, endorsed a statement including the following
sober warning:

The enormous world-wide spread of nuclear fissile material
(mainly plutonium) and of nuclear know-how, which is going to
occur in the next one or two decades to satisfy the energy demands
of the world, constitutes a problem of staggering proportions. It
is clear that the management of this problem will necessitate a

high degree of international collaboration, if disasters of major proportions are to be avoided. It is difficult to imagine that such an amount of collaboration will be possible unless detente and disarmament make substantial progress in the immediate future. There is a danger, to some degree already present, that processed fissile materials in storage or in transit may fall into the hands of irresponsible, possibly criminal or fanatical groups. The need for ensuring the physical protection of fissile materials, by both international and national means, must be strongly emphasized.

That there were such 'irresponsible, criminal or fanatical groups' at large was amply evident; it quickly became equally evident that they had not overlooked the possibility of nuclear malevolence. In Argentina a band of urban guerrillas took over the newly-constructed Atucha nuclear station, a 340-MWe pressurized heavy water reactor not far from Buenos Aires. The reactor had not yet gone critical; the guerrillas occupied the station for twenty-four hours, then left, having done nothing more injurious than spray-painting revolutionary slogans around the interior of the plant. On 12 November 1972 three thugs hijacked a DC-9 on a domestic flight up the US east coast. In the course of a tense two-day ordeal for the crew and passengers which reached as far as Toronto and Havana, the plane circled for two hours over the Oak Ridge AEC facilities in Tennessee, while the hijackers demanded $10 million, and threatened to crash the aircraft into the reactor installation. Oak Ridge officials did not consider the threat idle; they shut down all the reactors at the site and evacuated most of the staff. The plane had to land at Lexington, Kentucky, to refuel; when it took off renewed threats from the hijackers forced it to return to the Oak Ridge vicinity. They got their $10 million, plus bullet-proof jackets and pep pills, in an airport transaction at Chattanooga, Tennessee; then the DC-9 flew a second time to Havana, and was allowed to land by the Cuban authorities, to whom the hijackers surrendered. Oak Ridge officials said that an aircraft crashing on the reactors would at worst release a small amount of radioactivity to the immediate area. But the episode was far from reassuring.

Only a month later, on 14 December 1972, 1500 workers at the Dounreay research installation on the north Scottish coast were sent home after an anonymous telephone call warning that bombs had been planted on the site. Security staff found two suspicious parcels, one in the main workshop and the other at the entrance to the Dounreay fast reactor. The parcels were destroyed by an army bomb disposal squad, and proved to be harmless. But, once again, they might not have been. The recent history of bombings in Britain needs no reminder.

Clearly, the impulse towards nuclear malfeasance on every scale is abundantly present. What of the opportunities? Material like plutonium and highly enriched uranium from which nuclear weapons can be made is called 'special nuclear material', 'special fissile material', 'strategic material' or just 'SNM'. The April 1969 meeting of the Institute of Nuclear Materials Management heard Sam Edlow, a consultant on nuclear materials transport, relating a series of recent experiences, including some of his own. Strategic material in amounts sufficient for dozens of bombs was, according to Edlow, routinely lost, misrouted and overlooked by airlines, trucking companies and freight terminals. A shipment of his, thirty-three kilograms of 90 per cent enriched uranium travelling from New York to Frankfurt, was mistakenly offloaded at London Airport and left there unattended until the shippers asked the airline about it. A US domestic shipment from Ohio arrived in St Louis with one of three containers of strategic materials – gross weight 385 kilograms – missing. Not until nine days later did the missing container finally turn up – in Boston under a load of shoes.

By 1972 a substantial number of people within the nuclear community were becoming openly worried about the increasingly casual attitude towards strategic material. One of the most outspoken was a nuclear physicist named Ted Taylor, who during the 1950s was the AEC's star designer of fission bombs at Los Alamos. Taylor had been a contributor to the Kansas State symposium, as had Mason Willrich, a lawyer and one-time staff member of the US Arms Control and Disarmament Agency. In 1972 Taylor and Willrich were commissioned to prepare for the Ford Foundation's Energy Policy Project a study of nuclear theft. Their work occupied more than a year, during which Taylor visited many of the facilities in the USA, government and private, which shared responsibilities for the secure handling, transport and storage of special nuclear material. With him much of the time was a writer named John McPhee, who chronicled their colloquy in a remarkable book called *The Curve of Binding Energy*, first published as a three-part series in the *New Yorker* magazine in December 1973 (see Bibliography, pp. 293–4). The book could have been called 'alarmist', if there had not been at the same time two official reports drawing conclusions in broad factual agreement with McPhee's presentation. On 7 November 1973 the US General Accounting Office published its Report to the Congress on *Improvements Needed in the Program for the Protection of Special Nuclear Material* – which was a low-key title for a hair-raising document. The investigation looked into three out of ninety-five organizations licenced to possess strategic material in quantities sufficient to require compliance with AEC requirements for protection. Two of the sample of three failed signally to fulfil the requirements. Facilities had weak physical security barriers, ineffective guard patrols, ineffective alarm systems, inadequate automatic detection devices, and no action plan to deal with a theft of nuclear material. Investigators found that they could enter facilities undetected, climb over fences, pull fencing apart, cut through steel-panelled storage buildings with tin snips in minutes, reach access windows unobserved and almost unimpeded, and in general virtually help themselves as they wished.

What could they do with strategic material once they had it? Until fairly recently it had been dogma that both the necessary knowledge and the necessary technology were beyond the means of any but highly organized efforts – nothing short of a national government, and then only with a major national effort.

It was further assumed that plutonium would have to be made expressly for weapons use, that only 'weapons-grade plutonium' would produce an explosion. The reason given was that different isotopes of plutonium behave differently in a bomb. Plutonium-239 has a very low probability for spontaneous fission, and a high one for neutron-induced fission. But if plutonium-239 is left in a reactor some plutonium nuclei may absorb neutrons without undergoing fission, and become plutonium-240, and then plutonium-241 and -242. Plutonium-240 has a substantial probability for spontaneous fission. Accordingly, a sample of plutonium containing a sizeable fraction of plutonium-240 always contains a significant crossfire of neutrons from spontaneous fission of the 240. Conventional wisdom held that these neutrons would make a bomb made of such material go off prematurely, blowing itself apart before a comprehensive chain reaction could have a chance to take place. Since the higher isotopes of plutonium are to all intents and purposes impossible to separate from plutonium-239 it was felt that plutonium from commercial power reactors would be of little use to prospective bomb-makers.

This comforting thesis began to crumble by the early 1970s. US and European nuclear experts came to the conclusion that it might be difficult to predict the performance, the explosive yield of a bomb made of power-reactor plutonium – but that it would very probably go off, with an all too convincing result.

Willrich and Taylor left little doubt about the scale of the consequent problem. Their book, *Nuclear Theft: Risks and Safeguards* (see Bibliography, p. 293), was published in April 1974; it was a landmark study, an instant standard work on a blood-chilling subject. It summarized in dismayingly explicit detail all the information long since available in the open literature, analysed the types of potential bomb-material that the civil nuclear programme would generate, estimated quantities, identified categories of potential nuclear thieves, their motives and modes of operation – nations, political groups, criminal groups, terrorists, fanatics, a who's who of potential nuclear malefactors – and attempted to devise a coherent and feasible programme to repel any attempt at 'diversion' – the delicate industry euphemism which the uncompromising title of the Willrich–Taylor study scorned. Only the part of their book dealing with prevention failed to carry total conviction.

In the week-end of 26–30 April 1974, Senator Abraham Ribicoff disclosed to the press and to Congress that the AEC had carried out its own study of the potential misuse of nuclear materials. The AEC was apparently sitting on the report until it came into Ribicoff's hands, at which point AEC staff hastily passed out copies to the press, as if endeavouring to minimize the impact of Ribicoff's warning. But the report, known as the Rosenbaum report after Dr David Rosenbaum, one of its five authors, restated – in equally unambiguous language – the facts and the interpretations presented by Willrich and Taylor. It made clear that the security of nuclear materials had fallen far below the required standards – and indeed far behind the increasing production and circulation of such materials.

By this time the AEC was moving towards much more uncompromising conditions for management of nuclear materials. In a revision of its requirements under the Code of Federal Regulations, 10 CFR, the AEC called for armed guards on shipments of strategic material, and soon indicated that utilities should engage armed guards for nuclear power stations too. Neither of these provisions were at all palatable to the utilities, who were having a hard enough time persuading the public that nuclear power was a boon, without having to explain the presence of armed men around their facilities. In any case, as one utility spokesman noted, a nuclear power station is not a particularly suitable place for gunplay. However, in January 1974, retired Green Beret Colonel Aaron Bank told a hearing in San Diego that he could 'readily sabotage' the San Onofre nuclear power station, a 430-MWe pressurized water reactor on the Californian coast not far from Los Angeles. Colonel Bank's testimony has never been made public. But Joseph Schleimer, a reporter who interviewed Colonel Bank, later published in the *Bulletin of the Atomic Scientists* for October 1974 a terse description of the possibilities, which cannot have set southern Californian minds at rest.

In West Germany the news magazine *Der Spiegel* reported that its staff had succeeded in gaining access to the interior of a plutonium storage depot at Wolfgang near Hanau. The company operating the depot and the state government of Hesse declared that the reporters had not reached the area in which 300 kilograms of plutonium were stored; but the issue aroused public concern in West Germany. In Britain it was revealed that fuel from the Dounreay Prototype Fast Reactor would be reprocessed there, and that the recovered plutonium would then be shipped by road back to the plutonium plant at Windscale, a distance of nearly 600 kilometres; there would be about one shipment per month, containing perhaps 100 kilograms of plutonium. Officials declined to specify what security precautions would be taken, but were confident that they would be adequate. As is generally the case in Britain information about transportation or plant security is not on public record, neither the criteria applied nor any indication as to whether they are being effectively met in practice.

Then, as concern was mounting about the domestic security of fissile materials and nuclear facilities, came 18 May 1974, and the Indian nuclear explosion. Suddenly the nuclear community received a sharp reminder that not only terrorists and criminals might 'divert' fissile material for bombs – national governments, the originators of nuclear weapons, were still very much a major factor in the strategic materials problem. India was the first Third World country to demonstrate nuclear-weapons capability; but India was by no means the only possible candidate, Third World or otherwise. Ironically, the Indian bomb went off just as the authoritative Stockholm International Peace Research Institute was on the verge of publishing a study entitled *Nuclear Proliferation Problems*. The Institute's Yearbook, *World Armaments and Disarmament*, 1972 edition, had identified fifteen countries, which either had not signed or had signed but not ratified the Non-Proliferation Treaty, and whose nuclear capabilities might well be directed towards weapons. The countries included Argentina, Brazil, India, Israel, Pakistan, South Africa, and Spain – non-signatories – and Australia, Belgium, Egypt, Italy, Japan, the Netherlands, Switzerland and West Germany – then non-ratifiers.

The Indian move was far from unexpected; but it was none the less profoundly disturbing.

For its part India had never concealed its distaste for the Non-Proliferation Treaty. To India, the Treaty represented an attempt on the part of the nuclear weapons powers, especially the USA and the USSR, to calcify the *status quo*, to preserve their enhanced international status while in no way restricting the continuing growth of their own nuclear arsenals or committing them to any significant effort towards disarmament. In the Indian view the threat of nuclear proliferation at the level of national governments was miniscule compared to that posed by the nuclear activities of the major weapons-powers. Indian spokesmen pointed to the amount and availability of nuclear weapons materials in the USA; to the erratic behaviour of US military personnel in Vietnam and elsewhere; to the frequency and variety of criminal or fanatical exploits in the USA; and to the international link-ups between terrorists that might lead to the use of purloined US strategic material in some other part of the world.

Unfortunately, however valid such comments, they emphasize rather than offset the destabilizing effect of increased nuclear activity. The Canadian government was particularly upset by the Indian explosion. The reactor in which India manufactured the plutonium for her bomb was the CIRUS reactor at Trombay, a 40-MWt heavy water research reactor made with the help of Canadian scientists and engineers, during the long and vigorous cooperative programme carried on between India and Canada from the early 1950s onwards. The CIRUS reactor was not subject to International Atomic Energy Authority safeguards; such safeguards had not even been established when it first went critical in July 1960. But bilateral Canadian–Indian agreements led Canada to understand that no Canadian nuclear aid would be used by India to develop nuclear weapons. After the 18 May explosion the Indians simply declared that the device was not a weapon, but a peaceful explosive. The semantic distinction did not impress the angry Canadians, who forthwith cut off all further nuclear assistance to India. In due course other assistance, from sources with fewer scruples or at any rate a more flexible policy, was forthcoming, from France among others. India for her part almost immediately announced a bilateral arrangement to exchange nuclear know-how with Argentina, another Canadian customer, already building a CANDU power reactor at Rio Tercero, and in the market for another.

The effect of the Indian bomb on the Canadian nuclear export programme was traumatic. Just when the Canadian natural uranium reactor designs had found a place in the international market hitherto monopolized by the US light water designs, along came undeniable evidence that selling Canadian reactors to other countries might put them within reach of nuclear weapons. The Canadian government went into a brown study for months, wrestling with its conscience, while the Canadian nuclear salesmen stormed about waving their arms, lamenting the lost opportunities. Even the export of Canadian uranium, until this time a major economic part of the nuclear business in Canada, was subjected to agitated scrutiny with regard to its potentially less peaceful aspects.

US nuclear interests took a sanctimonious line that further infuriated the Canadians. US spokesmen referred repeatedly to the 'inadequate' safeguards on the CIRUS reactor – so unlike the scrupulous safeguards imposed on exported US reactors, they implied. Quite apart from the dangers of throwing stones in glasshouses, the US spokesmen overlooked a salient point: the heavy water inventory of the CIRUS reactor was supplied by the USA, whose complicity was accordingly immediate at whatever level of condemnation might be applicable. Meanwhile, to exacerbate matters yet further, President Nixon, fleeing the wrath to come, made his farewell tour of the Middle East, and offered nuclear largesse right and left. On 14 June it was reactors to Egypt, on 17 June reactors to Israel. The offers brought raised eyebrows elsewhere – and, in due course, in the USA itself, as the military aspects came more clearly into focus.

In Britain, while the condemnations of the Indian test and the US offer of reactors to the Middle East were still ringing out, rumours suddenly surfaced that Britain was on the verge of a nuclear test of her own. After a peevish silence Prime Minister Harold Wilson made a brief and – true to British nuclear form – unenlightening statement in the House of Commons on 24 June, disclosing that Britain was not about to carry out a nuclear test. Britain had in fact already done so, in the second week in June, at the Nevada Test Site in the USA.

The *Bulletin of the Atomic Scientists*, for nearly three decades an uncompromising and committed commentator on the global implications of nuclear technology and policy, has always carried on its title page a succinct image portraying the state of nuclear affairs. It is the top left-hand quadrant of a clock. The hand of the clock is approaching midnight. When nuclear circumstances dictate, the hand of the *Bulletin* clock moves. At the time of the Cuban missile crisis in 1962 the *Bulletin* clock advanced to three minutes to midnight. By the time of the first Strategic Arms Limitations Talks in 1972 the clock had been set back to ten minutes to midnight. The first Strategic Arms Limitations agreement moved it back yet another two minutes to twelve minutes to midnight. After the Indian bomb, reactor offers to the Middle East, and burgeoning concern about nuclear security at every level, the Bulletin clock stopped retreating. In September 1974 it advanced to nine minutes to midnight. On 4 December 1974 Israel announced unequivocally that she could manufacture nuclear weapons at will.

10. The Nuclear Horizon

Take a deep breath. Now: by 31 December 1974 there were fifty-three nuclear power reactors within the USA licensed to operate, with a capacity of nearly 39 000 MWe. There were twenty-nine power reactors in operation in the UK, with a capacity of 5600 MWe. There were sixteen power reactors in operation in the USSR, with a capacity of 3000 MWe. There were ten power reactors in operation in France, with a capacity of nearly 3000 MWe. There were power reactors in operation also in Canada, Czechoslovakia, East Germany, India, Italy, Japan, the Netherlands, Pakistan, Spain, Sweden, Switzerland and West Germany. In all, there were 149 operable power reactors with a total capacity of some 58 000 MWe. There were power reactors under construction or on order also in Argentina, Austria, Belgium, Brazil, Bulgaria, Finland, Hungary, Korea, Mexico, the Philippines, Taiwan, Thailand and Yugoslavia. There were also power reactors planned but not yet ordered in Iran, Israel, Portugal and South Africa. The total number of all these reactors – in operation, under construction, ordered or planned – was 401 with a total output of some 255 000 MWe. Nine other countries – Bangladesh, Chile, Denmark, Egypt, Greece, Jamaica, Romania, Singapore and Turkey – have announced plans to build power reactors by the year 2000. Most of the above countries and some others already possess research reactors; some of these, in for example Israel, Norway, Poland, South Africa and Yugoslavia, are comparable in heat output – and plutonium production – to small power reactors.

Many of the above countries also have, or plan to acquire, some if not all of the other components of the nuclear fuel cycle: uranium mines and mills, hexafluoride plants, enrichment plants, fuel fabrication plants, fuel-reprocessing plants, and waste management facilities. Six countries are known to possess

Table 3   The growth of nuclear power

The figures are for number of power reactors in operation, and the total installed nuclear electricity capacity (MWe) at the end of each year. Figures from 1954 to 1974 are actual, from 1975 to 1980 estimated

Source: IAEA

	Number of reactors	Installed capacity
1954	2	7
1955	2	7
1956	6	113
1957	10	214
1958	18	753
1959	22	1114
1960	24	1304
1961	27	1569
1962	41	2895
1963	54	4568
1964	62	6100
1965	66	7106
1966	74	8483
1967	80	10172
1968	81	10894
1969	89	14942
1970	98	20015
1971	111	26166
1972	128	34530
1973	149	47162
1974	182	72417
1975	219	100058
1976	251	127081
1977	278	149008
1978	303	170691
1979	347	212435
1980	401	262896

nuclear weapons technology; others are not far behind.

Governmental and industrial enthusiasm for nuclear power – in both senses of the expression – is burgeoning rapidly. The rise since 1970 in the price of crude petroleum has been taken as a sign that future energy supplies must include a large and growing proportion of nuclear energy. Projections by the International Atomic Energy Agency, the Organization for Economic Cooperation and Development, the European Economic Community, the US Atomic Energy Commission, the UK Atomic Energy Authority, the French Commissariat d'Energie Atomique and many other organizations and individuals in the nuclear field assert that nuclear energy will supply 50 per cent or more of world electricity requirements by the year 2000. Estimates of world-wide nuclear generating capacity in 2000 have ranged at least as high as 4 500 000 MWe. – that is, 4500 power reactors of output as high as the largest now operating.

From 1970 onwards the nuclear option began to play a major role in general energy policy, particularly among the industrialized nations. The desire to reduce dependence on the petroleum-exporting countries, especially those of the Middle East, manifested itself in a determination on the part of the

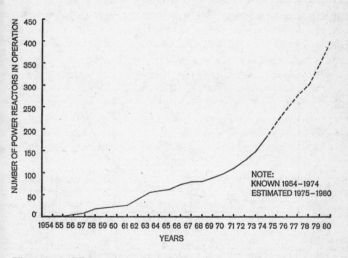

Figure 12   The growth of nuclear power: number of operating reactors

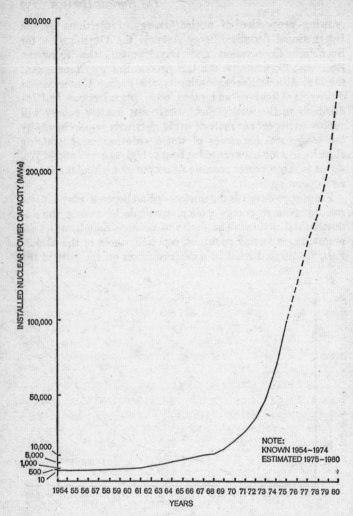

Figure 13, The growth of nuclear power: installed nuclear electricity capacity

USA, France, Japan and other vulnerable Western countries to expand their nuclear electrical generating capacity as rapidly as possible. The marginal cost advantage of electricity from nuclear stations was dramatically enhanced by the oil price rise, coupled with the renewed insistence of coal-miners in Britain, the USA and elsewhere on hefty pay rises. Attempts to boost coal output were further hampered by the prolonged economic shakiness of the industry in Britain and France, and a bitter battle about air quality standards and the effects of strip mining in the USA. Nuclear energy seemed by comparison a manageable technology, ready to expand to meet the prognostications of future demand with minimal effect on the environment and on the health of its workers. Concern about a possible imminent shortage of uranium could be set aside by referring to the forthcoming role of the fast breeder reactor, improving fifty-fold the utilization of uranium and – in a country like Britain – even suggesting that the import of uranium would be unnecessary for the indefinite future.

By such a time, it was anticipated, other nuclear energy technologies would be playing their part. Nuclear-powered ships, including cargo submarines, tankers and bulk carriers would be able to cruise the oceans at will, with no fear for fuel costs. Nuclear generating stations would be built, not merely on remote coastlines, but in near-urban locations, where the heat output from their turbines, formerly wasted, would supply district heating to entire neighbourhoods and industrial estates, incidentally bringing the fuel costs yet lower, and boosting efficiencies to unheard-of heights. Meanwhile new reactor designs, developments of the high-temperature reactor, would take over from the dwindling supplies of fossil fuel, and supply process heat for industry; nuclear steel-making seemed likely to be the first breakthrough on this front. As a demonstration of the ways in which nuclear energy would help to alleviate other resource-problems, desalination of sea-water using the low-temperature heat from reactors was expected to become an important contribution to water supplies. Desalination was a service expected to be especially welcome in some developing countries, whose burgeoning energy requirements would be impossible to fulfil from remaining fossil reserves, leaving only the nuclear avenue to development.

Such a view has an undeniable appeal. But it sidesteps a number of knotty questions, to which answers are not as yet fully forthcoming. On a purely economic level the problem of building nuclear capacity at desired rates and of desired magnitudes is proving formidable. Over the last century the trend in energy supply, from wood to coal to oil to gas, has always moved in the direction of inherently simpler technologies and sources. The nuclear option is an abrupt departure from this trend. It is doubtful whether it can be introduced quickly enough both to take over from previous sources, and to sustain envisioned rates of growth. Major programmes like those of the USA and France are finding themselves up against shortages of money, of material resources, and of adequately trained manpower.

Nuclear power is not only heavily capital-intensive, but also energy-intensive: that is, it requires a great deal of energy to build and to service nuclear power stations. Recent work in the new discipline of 'energy accountancy' or 'dynamic energy analysis', has suggested some disquieting aspects of the energy throughput required by major expansion programmes. A single nuclear station, during its operating life-time, should produce substantially more energy than has been consumed in its construction and fuelling. But, according to work at the British Open University and elsewhere, an exponentially increasing programme of nuclear stations would require a large proportion of the output from stations already operating simply to support work on those under construction. If the programme were to expand rapidly enough – like that of France, which is said to double every two years – stations in operation might be unable to supply the energy requirements for those still under construction. During this period of growth, which might last ten years or more, the economy of the country could find itself carrying an extra burden on behalf of its nuclear capacity, at a time when its energy position is already precarious; such seems to be the case with France at the time of writing.

To be sure, when the growth phase ends, the newly operational nuclear capacity will be a major addition to energy supply. But some analysts believe that its availability may more or less coincide with the sharp fall-off in energy demand caused by the reduction in nuclear construction activities. The country's energy-supply position may therefore swing rapidly from famine to glut – with possibly disruptive social and economic consequences.

Other energy supply technologies may exhibit similar peculiarities; investigations are still in progress. Hard data on nuclear energetics are scarce, and some of the analysis described above is still hotly disputed. But it seems likely that net energy considerations – how much useful energy is produced in return for how much energy invested – will become an additional significant factor in energy policy planning, especially as it affects energy-intensive production technologies like nuclear electricity.

We have alluded earlier to the hazards arising from the radioactivity of the materials of the nuclear fuel cycle. The effects of low-level radiation, and the gradual build-up of manmade environmental radioactivity, are acutely difficult to ascertain; if, in due course, they are found to be a serious liability to global ecology and to the genetic vigour of living things, it may well be too late to remedy the situation. The same considerations apply, of course, to other man-made contaminants like heavy metals and persistent chemicals; the nuclear worry is not unique except insofar as we know already the genetic vulnerability of organisms to radiation, and do not know how to extrapolate to an entire planet or to many generations of human exposure.

A more acute problem of a similar nature arises from the accumulation of high-level radioactive wastes from fuel reprocessing. Fission products like strontium-90 and caesium-137 remain dangerous for hundreds of years, actinides like plutonium-239 and americium-241 for hundreds of thousands of years. No human artefact can guarantee to isolate such substances for such a period. The volumes of such wastes are, to be sure, small by comparison with wastes from some other energy technologies – pit spoil and ash from coal-fired stations, or sulphurous sludge from stack-gas scrubbers, for instance. If high-level wastes can be solidified, into tough borosilicate glass, their mobility can be minimized. But the questions remain – not so much technical as ethical. Are we justified in availing ourselves of nuclear energy, if by so doing we impose an effectively permanent burden on our descendants? On the other hand, are we justified in confining our energy-requirements to fossil fuels, and thereby depleting them to a pittance for future generations?

More immediate problems also arise. The nuclear fuel cycle abounds with complex technology, in installations which contain awesome amounts of radioactive materials. What assurance have we that such installations can be guaranteed to operate safely? A breach of containment, either by accident or by sabotage, could release enough radioactivity to render a vast area indefinitely uninhabitable. Some analysts, notably those who drafted the AEC's WASH-1400 Reactor Safety Study, have concluded that the probability of a major accident to a light water reactor is minuscule. But their conclusions have not gone unchallenged; and there remain the questions of other reactor designs, of other fuel-cycle installations like reprocessing plants, and – unfortunately – of sabotage.

The most dismaying problem of all is probably that of fissile material security: guaranteeing that potential nuclear weapons material – plutonium-239, uranium-233 and uranium-235 – does not fall into the wrong hands, whether of trigger-happy governments or of terrorists. The possibility is far from hypothetical, and does not seem amenable to easy solution. US nuclear interests have proposed the establishment of a National Fissile Material Security Service – a federal agency like the FBI and the CIA, responsible for overt and covert supervision of the nuclear fuel cycle. It is not an attractive idea, indicating a trend towards authoritarian centralized control of a society, of the kind most likely to precipitate just the social unrest which would most readily lead to nuclear upheaval. It also prompts mention of William T. Riley, former top security officer of the AEC, who was sentenced in February 1973 to three years' probation for having borrowed $239 000 from fellow AEC employees, and failing to pay back over $170 000, having used most of the money for betting on horse races. Who, in a Nuclear Security Service, will watch over the custodians?

The problem, furthermore, is not confined within national boundaries. Laxity at key points anywhere in the world may have repercussions anywhere in the world. The ramifications of international terrorism need no recapitulation. In this light the enthusiasm for export of reactor technology from the USA, the UK, France, Canada, and the Soviet Union to other countries, some of doubtful stability, seems almost perverse.

Fred Iklé, director of the US Arms Control and Disarmament Agency, speaking in January 1975, pointed out a grimly sobering aspect of the situation. With nuclear technology and materials ever more widespread, what would happen if a nuclear explosion wiped out, say, Washington? The USA might have no firm idea who had done it – and thus the entire monolithic theory of strategic mutual deterrence crumbles into blatant ineffectuality. Curiously enough, at a time when access to resources seems likely to provoke one showdown after another, the widespread dissemination of nuclear technology will undoubtedly have an unexpected, if discomfiting, equalizing effect. No matter how small a nation, no matter how small a group of participants, if they have access to nuclear weapons, or even to quantities of long-lived radioactivity, their voice must be heard with respect. From this point of view a country with many vulnerable nuclear installations is offering hostages to all and sundry.

Large power stations – whether nuclear or fossil-fuel – and their related electrical grid systems are vulnerable in a number of ways, both during construction and when in operation. Nuclear power stations cannot readily be built economically below a

certain size, at least 100 MWe, and can easily be shown to be
more economic an order of magnitude larger. Unfortunately
stations of such a size take a long time to build – at least five
years, and in some cases as many as ten. This means that such a
station is being built in a planning vacuum; credible forecasts
of electrical demand more than five years ahead can no longer
be made. Stations of such a size are prone to engineering problems of many kinds, during construction and during operation;
and a small malfunction may make a 1000-MWe station unavailable for months, at prohibitive cost to the operator. Precisely such a coincidence of effects arose in Britain in the
1960s. On the basis of a forecast 8 per cent growth in electricity
demand the Central Electricity Generating Board ordered a
vast programme of new capacity, including the ill-fated
advanced gas-cooled reactor stations. These stations and the
forty-seven very large new turbogenerator sets, each of 500
MWe, which were used in both nuclear and fossil-fuel stations
caused seemingly endless problems. Fortunately for the
CEGB, the forecast growth rate in demand was also drastically
inaccurate; the actual growth rate was less than 3 per cent, and
the unavailable stations were not needed anyway. The losers
were the electricity consumers and the taxpayers, who had to
finance the unnecessary investment in new capacity, through
electricity rates and taxes.

Many commentators have begun to question the alleged
economies of scale of very large power stations. Economies may
indeed appear on paper; but if construction delays, unforeseen
engineering problems, and subsequent breakdowns are allowed
for in calculations the economic size of stations seems much
smaller than those now being ordered. Furthermore, a power
station producing 1000 MW of electricity also produces perhaps 2000 MW of heat – far too much to be useable even in the
largest industrial installation. So such giant stations have to be
sited at a remote location, where the heat can be discharged to
the surroundings as waste – usually in wilderness or amenity
areas, which must be further disrupted to provide transmission
facilities, which in turn entail further losses of useful energy.

Accordingly, attention is now turning to different philosophies and technologies of energy supply. The majority of energy required in an industrial economy is not high-grade or high-temperature energy like electricity but low-grade heat. In some countries, like Japan, Australia and parts of the southern USA, solar energy has long been used to heat and cool buildings and provide hot-water supplies; interest in low-temperature solar energy applications is now burgeoning world-wide. Even in higher latitudes it is being realized that solar energy can provide a useful input to the total energy supply, even if only for the first stage of water heating. Capital costs are still high, but are expected to come down as the technology matures and acquires more widespread applications. Wind, too, is being recognized as a possible contributor to energy supply in some regions. Like direct solar energy, wind is an inherently decentralized source of energy; both direct solar energy and wind may make an economically significant contribution to energy supply, as the advantages of decentralized sources are recognized.

Geothermal energy – energy from the hot interior of the earth's crust – is also under investigation, as a source both of direct heat and of steam to drive turbogenerators. Its availability will vary from region to region, but may be substantial. Biological processes for generating fuels like methane from organic waste materials have long been established in some localities; as the prices and availabilities of the fossil supplies become less favourable biogeneration may play a newly economic role. More exotic technologies have also been proposed, including offshore structures to tap the energy of ocean waves, and to utilize the temperature gradients between the deep ocean and its surface.

Advocates of these alternative energy sources do not claim that any one will provide for all requirements. They say rather that a mix of different sources, matching demand in quality as well as quantity, is feasible and achievable within present constraints of finances, resources and time. They point to the funding of energy research and development, which in the past twenty-five years has been concentrated on nuclear energy almost to the exclusion of other technologies, even those based on fossil fuels, like coal gasification and liquefaction. If even a modicum of the available research and development effort were redirected into the alternatives, they feel that the nuclear option would soon appear neither the only option nor the best. Others disagree; in their view only the rapid development of nuclear technology will supply the energy needed by the people of the earth.

We stand today before an abundance of potentials and possibilities; the options are still open. Within the present generation they will almost certainly be foreclosed. The decisions now impending will affect not merely energy supply and demand, but the entire organization of our global society. We, the people of the world, must be party to these decisions. Before we commit ourselves and our descendants to a nuclear future, it is vital that we concur in and understand the nature of the commitment. If we undertake it now we do so for all time.

Appendix A

Nuclear Jargon

ABCC: Atomic Bomb Casualty Commission; US organization in Japan responsible for victims of Hiroshima and Nagasaki bombs.

ACRS: Advisory Committee on Reactor Safeguards; responsible for assessment of safety of reactors licensed in the USA.

Actinide: one of the heavy elements actinium, thorium, protactinium, uranium, neptunium, plutonium, americium, curium, berkelium and californium, all of which are chemically very similar; actinides of interest are those which are long half-life alpha-emitters.

Activation: absorption of neutrons to make a substance radioactive.

AEA, UKAEA: United Kingdom Atomic Energy Authority.

AEC, USAEC: United States Atomic Energy Commission.

AECL: Atomic Energy of Canada Ltd.

AGR: advanced gas-cooled reactor.

Alpha Particle: high-energy helium nucleus (two protons, two neutrons) emitted by some heavy radioactive nuclei.

Atom: see pp. 23–4.

Beta Particle: high-energy electron emitted by radioactive nucleus.

Boron: powerful absorber of neutrons used – usually in alloy steel – for reactor control rods et cetera.

Breeder: reactor which produces more fissile nuclei than it consumes.

Breeding Gain: proportional increase in fissile nuclei in fuel after its removal from breeder.

Burner: reactor which consumes more fissile nuclei than it produces.

Burn-Up: cumulative output of heat from reactor fuel; directly correlated with build-up of fission products; usually measured in megawatt-days per tonne of uranium.

Butex: organic solvent used in reprocessing irradiated reactor fuel.
BWR: boiling water reactor.

Caesium: particularly caesium-137; fission product, biologically hazardous beta-emitter.

Calandria: in pressure-tube reactor designs, tank containing moderator – usually heavy water – and through which run pressure tubes.

CANDU: Canadian Deuterium Uranium reactor.

Cave: room with heavily shielded walls, within which highly radioactive materials can be handled by remote control.

CEA: Commissariat d'Energie Atomique (France).

CEGB: Central Electricity Generating Board (UK).

Cerenkov Radiation: blue light emitted when nuclear radiation travels through a transparent medium (like water) at a speed greater than that of light in the medium.

Chain Reaction: see pp. 30–32.

China Syndrome: possible consequence of core meltdown, when a molten mass of intensely radioactive material plummets through vessel and containment and into the earth beneath in the direction of China (unless the reactor is in, say, Japan).

Cladding, sometimes just *Clad* (as noun): metal sheath (Magnox, zircaloy, stainless steel or ceramic) within which reactor fuel is hermetically sealed.

CND: Campaign for Nuclear Disarmament (UK).

CNI: Committee for Nuclear Information (USA).

Containment: structure within a reactor building – or the building itself – which acts as a barrier to contain any radioactivity which may escape from the reactor itself.

Contamination: radioactivity where it should not be.

Control Rod: rod of neutron-absorbing material inserted into reactor core to soak up neutrons and shut off or reduce rate of fission reaction.

Conversion Ratio: number of fertile nuclei converted to fissile, compared to number of fissile nuclei lost by undergoing fission.

Coolant: liquid (water, molten metal) or gas (carbon dioxide, helium, air) pumped through reactor core to remove heat generated in the core.

Cooling Pond: deep tank of water into which irradiated fuel is discharged upon removal from a reactor, there to remain until shipped for reprocessing.

Core: the region of a reactor containing fuel (and moderator, if any) within which the fission reaction is occurring.

Critical: refers to a chain reaction in which the total number of neutrons in one 'generation' of a chain reaction is the same as the total number of neutrons in the next 'generation' of the chain; that is, a system in which the neutron density is neither increasing nor decreasing.

Criticality: the state of being critical.

Criticality Accident: inadvertent accumulation of fissile material into a critical assembly, accompanied by outburst of neutrons and gamma radiation.

Cross-Section: hypothetical 'target-area' measuring the probability of a nuclear event.

Crud: impurity deposit inside a reactor.

Curie: amount of radioactive material giving off 37 000 million radioactive emissions per second; radioactivity of 1 gram of radium.

Daughter, or *Daughter Product*: the substance into which a radioactive nucleus transforms itself by radioactive decay.

Decay: radioactive transformation.

Decay Heat: heat generated by radioactivity in the fuel of an operating reactor; additional to heat from chain reaction, and cannot be shut off.

Decontamination: transfer of unwanted radioactivity to a less undesirable location.

Densification: compaction of fuel inside cladding, as a consequence of irradiation; can lead to fuel damage because of unbalanced internal and external pressures.

Depleted Uranium: uranium with less than the natural proportion (0.7 per cent) of uranium-235, which has been removed in an enrichment process and transferred to the remaining 'enriched' uranium.

Deuterium: hydrogen-2, heavy hydrogen; its nucleus consists of one proton plus one neutron, rather than the one proton only of ordinary hydrogen.

Deuterium Oxide, heavy water: water in which the hydrogen atoms are heavy hydrogen.

Deuteron: nucleus of heavy hydrogen.

Disassembly: structural damage within a reactor core as a result of an excessive release of energy; 'blowing apart'.

Divergence: the achievement of criticality; 'going critical'.

Diversion: euphemism for theft, as applied to strategic or 'special' nuclear material.

Dose: amount of energy delivered to a unit mass of a material by radiation travelling through it.

Dose-Rate: time rate at which radiation delivers energy to unit mass of a material through which it is travelling.

Doubling Time: time taken for a breeder reactor to produce additional fissile material enough to duplicate its total 'pipeline inventory' (see p. 85).

Drywell: on a boiling water reactor, the concrete containment around the reactor pressure vessel.

ECCS: emergency core cooling systems.

Electron: negatively charged particle; much lighter than proton or neutron.

Enriched, as in *Enriched Uranium*: uranium in which the proportion of uranium-235 has been increased above the natural 0.7 per cent.

Enrichment: process of making enriched uranium.

ERDA: Energy Research and Development Administration; one of the two US Federal agencies created after the split-up of the AEC.

Excited: having an excess of energy.

Exposure: to radiation: passage of radiation through a material.

Fallout: radioactive fission products created by nuclear explosions, which descend from the atmosphere onto the surface of the earth.

Fast: of neutron: high energy, direct from fission.

Fast Breeder: reactor designed to have conversion ratio greater than 1, using unmoderated fast neutrons.

FBR: fast breeder reactor.

Fertile: of material like uranium-238 or thorium-232, which can by neutron absorption be transformed into fissile material.

Fissile: capable of undergoing fission.

Fission: rupture of a nucleus into two lighter fragments (*Fission Products*) plus free neutrons – either spontaneously or as a consequence of absorption of a neutron.

Flux: of neutrons, moving cloud of particles, particularly in reactor core: number of neutrons through unit area in unit time.

FOE: Friends of the Earth.

Fuel: material (such as natural or enriched uranium or uranium and/or plutonium dioxide) containing fissile nuclei, fabricated into a suitable form for use in a reactor core.

Fuel Assembly, *Fuel Element*: single unit of fuel-plus-cladding which can be individually inserted into or removed from reactor core.

Fuel Pin: single tube of cladding filled with pellets of fuel.

Fuel Rating: instantaneous power output per unit mass of fuel; measured as kilowatts per kilogram of uranium; also known as *Specific Power*.

Fusion: the combination of two light nuclei to form a single heavier nucleus.

Gamma Ray: high-energy electromagnetic radiation of great penetrating power emitted by nucleus.

Gas Centrifuge: uranium enrichment device by which heavier uranium-238 nuclei are slightly separated from lighter uranium-235 nuclei by centrifuging of uranium hexafluoride gas; full-scale plant uses many thousands of centrifuges in cascade.

Gaseous Diffusion: uranium enrichment process utilizing slight difference in rate of diffusion of uranium-235 and -238 hexafluoride molecules through porous metallic membrane; full-scale plant uses many thousands of diffusion cells in cascade.

Gigawatt: one thousand million watts.

Graphite: black compacted crystalline carbon, used as neutron moderator and reflector in reactor cores.

GS (*Girdler–Sulphide*): process used for the production of heavy water.

Half-Life: period of time within which half the nuclei in a sample of radioactive material undergo decay; characteristic constant for each particular species of nucleus.

Heat Exchanger: boiler, in which hot coolant from reactor core raises steam to drive turbogenerator; see also *Intermediate Heat Exchanger*.

T–L

Heavy Hydrogen, Heavy Water: see *Deuterium, Deuterium Oxide*.

Helium: light chemically inert gas used as coolant in high temperature reactors.

Hex: uranium hexafluoride, easily vaporized uranium compound used in enrichment processes.

High-Level: of radioactive waste, intensely radioactive with medium to long half-life.

Hot Cell: see *Cave*.

HTGR: high temperature gas-cooled reactor.

IAEA: International Atomic Energy Agency.

ICRP: International Commission on Radiological Protection.

Intermediate Heat Exchanger: tube array in a sodium-cooled reactor in which hot radioactive primary sodium coolant transfers heat to non-radioactive secondary sodium coolant.

Iodine: as iodine-131; biologically hazardous fission product of short half-life (8 days) which tends to accumulate in the thyroid gland.

Ion: atom shorn of one or more electrons, and therefore electrically charged.

Ionizing Radiation: radiation which can deliver energy in a form capable of knocking electrons off atoms, turning them into ions.

Irradiated: of reactor fuel, having been involved in a chain reaction, and having thereby accumulated fission products; in any application, exposed to radiation.

Isotope: form of an element, with the same number of protons in its nucleus as all other varieties of the element, but a different number of neutrons from other varieties of the element.

JCAE: Joint Committee on Atomic Energy of the US Congress.

Kilowatt: one thousand watts.

Krypton: a chemically inert gas; the isotope krypton-858 is a troublesome fission product at present released to the atmosphere from reprocessing plants.

Laser Enrichment: separation of uranium-235 from -238 selective excitation of one isotope with a laser; potentially a short-cut to highly-enriched uranium, which would present a serious problem as regards possible misuse of fissile material.

Light Water: ordinary water – to distinguish it from heavy water.

LMFBR: liquid metal fast breeder reactor.

Load Following: varying the power level of a reactor to match requirements of an electricity distribution system.

LOCA: loss-of-coolant accident.

Low-Level: of radioactive waste, not particularly radioactive.

LWR: light water reactor – either pressurized water reactor or boiling water reactor.

Magnox: alloy used as fuel cladding in first-generation British gas-cooled reactors, which are therefore called Magnox reactors.

Manhattan Project: the 'Manhattan District' of the US Army Corps of Engineers – code-name for the project which developed the atomic bomb.

McMahon Act: The Atomic Energy Act 1946, which banned any further transfer of nuclear information from the USA to the ertswhile allies Britain and Canada, and set up the US Atomic Energy Commission (AEC) and the Joint Congressional Committee on Atomic Energy (JCAE).

Megawatt: one million watts.

Meltdown: of reactor core, consequence of overheating which allows part or all of the solid fuel in a reactor to reach the temperature at which cladding and possibly fuel and support structure liquefy and collapse.

Mixed Oxide: of reactor fuel, fuel in which the fissile nuclei are plutonium-239, mixed with natural or depleted uranium in a proportion equivalent to enriched uranium.

Moderator: material whose nuclei are predominantly of low atomic weight (e.g. light water, heavy water, graphite) used in reactor core to slow down fast neutrons to increase probability of their absorption in uranium-235 or plutonium-239 to cause fission.

MUF: material unaccounted for; refers to discrepancy between amount of fissile material expected at any point in the fuel cycle and amount actually measured; may indicate that *diversion* has occurred.

MWE: megawatts electric.

MWT: megawatts thermal.

NAK: sodium–potassium alloy with low melting point, used as

coolant in early fast breeder reactors and as emergency coolant in some later designs.

Neutron: uncharged particle, constituent of nucleus – ejected at high energy during fission, capable of being absorbed in another nucleus and bringing about further fission or radioactive behaviour.

NPT: Non-Proliferation Treaty, intended to control the spread of nuclear weapons and their technology.

NRC: Nuclear Regulatory Commission, successor to the AEC with responsibility for licensing nuclear facilities in the USA.

NRDC: Natural Resources Defense Council (USA).

NRPB: National Radiological Protection Board (UK).

NSSS: nuclear steam supply system – in a nuclear power station, everything up to but not including the turbogenerators: reactor and its facilities (refuelling machine, control installation, fuel handling bay, cooling pond, steam generators if applicable, et cetera).

Nuclear Reactor: see pp 32–4.

Nucleon: either proton or neutron.

Nucleus: see pp. 23–4.

Nuclide: nucleus of isotope; nuclear species.

Off Gas: radioactive gas from within a reactor which is released to the atmosphere, usually after a delay to reduce its radio-activity.

Period: of reactor, time taken for a certain increase (or decrease) in power level; short period makes a reactor difficult to control.

Pile: formerly, nuclear reactor – after the first reactor, Chicago Pile No. 1.

Pipeline Inventory: total amount of fissile material associated with one reactor: amount in operating core, in cooling pond, in reprocessing plant, in fuel fabrication plant and in transit.

Plowshare: US programme for civil engineering applications of nuclear explosives.

Plutonium: heavy artificial metal, made by neutron bombardment of uranium; fissile, highly reactive chemically, extremely toxic alpha-emitter.

Power Density: in a reactor core, heat output per unit volume; measured in kilowatts per litre.

Pressure Suppression Pool: in a boiling water reactor, circular tunnel at bottom of *drywell*, half-filled with water, to condense steam from reactor cooling system if necessary.

Pressure Vessel: large container of welded steel or prestressed concrete within which are reactor core and other reactor internals.

Pressurizer: in a pressurized water reactor, electrically heated boiler in cooling system which boils water as necessary to maintain coolant pressure.

Price–Anderson Act: US Act of Congress limiting the third-party insurance liability of reactor operators in the event of an accident, and providing Federal indemnity to this limit.

Proton: positively charged particle, constituent of nucleus.

PTB: Partial Test Ban – treaty banning tests of nuclear weapons in the atmosphere.

Purex: organic solvent used in reprocessing of irradiated reactor fuel.

PWR: pressurized water reactor.

QA: Quality Assurance.

Rad: radiation absorbed dose; measure of exposure to radiation.

Radiation, Nuclear: neutrons, alpha or beta particles or gamma rays which radiate out from radioactive substance.

Radioactivity: behaviour of substance in which nuclei are undergoing transformation and emitting radiation; note that radioactivity produces radiation – the two terms are *not* equivalent.

Radiogenic: caused by radiation, as certain types of disease.

Radioisotope: radioactive isotope.

Radionuclide: radioactive nuclide.

Radium: intensely radioactive alpha-emitting heavy element.

Radon: alpha-emitting radioactive gas given off by radium.

Reactivity: measure of ability of assembly of fissile material to support sustained chain reaction. *Coefficient of Reactivity*, measure of the way the reactivity of an assembly changes in response to any other change, as for instance of temperature.

Reflector: of neutrons, a material of low atomic weight (light or heavy water, graphite) around a reactor core to reflect neutrons back into the reaction region.

Refuelling: replacement of reactor fuel after it has sustained maximum feasible *burn-up*; necessitated by loss of *reactivity*, build-up of neutron-absorbing fission products, and cumulative damage from radiation, temperature, coolant et cetera.

REM: Roentgen equivalent man: unit of radiation exposure, compensated to allow for extra biological damage by alpha particles or fast neutrons.

Reprocessing: mechanical and chemical treatment of irradiated fuel to remove fission products and recover fissile material.

Runaway: accidentally uncontrolled chain reaction.

Running Release: planned emission of radioactive material to the outside air or water.

Safeguards: term applied to keeping track of special nuclear material to prevent *diversion*.

Scram: emergency shutdown of fission reaction in reactor.

Separative Work: measure of energy required to enrich uranium.

SGHWR: steam generating heavy water reactor.

Shielding: wall of material (concrete, lead, water) surrounding source of radiation, to reduce its intensity.

SIPI: Scientists' Institute for Public Information (USA).

SNM, *Special Nuclear Material*: fissile material potentially useable in nuclear weapons.

Specific Activity: radioactivity per unit mass.

Specific Power: heat output per unit mass of fuel; see *Fuel Rating*.

Steam Generator: boiler, in which hot coolant from reactor raises steam to drive turbogenerator.

Strontium: isotopes, particularly strontium-90; fission products, biologically hazardous beta-emitters.

Sub-Critical: insufficiently supplied with neutrons to sustain a self-propagating chain reaction.

Tailings: fine grey sand, left over from extraction of uranium from ore; it contains radium, emits radon.

Tails Assay: amount of fissile uranium-235 left in uranium depleted during enrichment process.

Thorium: fertile heavy metal.

Tritium: hydrogen-3 – nucleus contains one proton plus two neutrons; radioactive.

Uranium: heaviest natural element, dark grey metal; isotopes 233 and 235 are fissile, 238 fertile; alpha-emitter.

USW: unit of separative work (p. 93).

Vitrification: fusing of high-level waste into glass-like solid.

WASH-740: AEC document *Theoretical Possibilities and Consequences of Major Accidents in Nuclear Power Plants* (1957).

WASH-1250: AEC document *The Safety of Nuclear Power Plants (Light Water Cooled) and Related Facilities* (1973).

WASH-1400: AEC document *An Assessment of Accident Risks in US Commercial Nuclear Power Plants* (1974).

Watt: measure of rate of transfer of energy; an adult human being gives off between 100 and 200 watts of heat.

Wigner Energy: energy stored in graphite moderator as a result of deformation by radiation.

Xenon Poisoning: accumulation of neutron-hungry fission product xenon-135, reducing reactivity of reactor.

Yellow Cake: mixed uranium oxides, with formula U_3O_8, produced from uranium ore by extraction process in uranium mill.

Zircaloy: alloy of zirconium used as fuel cladding; has low cross-section for absorption of neutrons.

Appendix B

Ionizing Radiation and Life

In Chapter 1 (pp. 25–8) we described how nuclear activities can produce four types of 'ionizing radiation': alpha, beta and gamma radiation and neutrons. There and elsewhere we indicated briefly some of the evidence accumulated since the discovery of radio-activity, about the effects of ionizing radiation on living organisms including human beings. The study of these effects is called 'radiobiology'. It is a subject of deep and controversial complexity, the more so since man began to create radioactivity in quantity. It is far beyond the scope of this book to describe in detail the findings of radiobiology. However, since most of the immediate potential nuclear hazards arise from the effects of ionizing radiation on living things some radiobiology is essential to pinpoint contentious issues.

In what follows 'radiation' means 'ionizing radiation' – not, for instance, sunlight (see pp. 28–30).

It is agreed that, fundamentally, ionizing radiation is not good for you. The passage of alpha, beta or gamma radiation, or neutrons through living tissue transfers energy to the atoms and molecules of the tissue, in a way which is bound to be more or less disruptive to the delicate organization of living systems. The disruptive effect is roughly proportional to the 'linear energy transfer' of the radiation. Beta and gamma radiation are of low linear energy transfer, alpha and neutron radiation of high linear energy transfer – also depending on the energy of the radiation. However, unlike a bullet in the brain, radiation – except in massive doses – is comparatively subtle in its effects. The radiation is invisible, and so, in almost all cases, is the damage. However, after radiation energy has disrupted some molecules in a living cell, the cell's biochemical behaviour may be affected. Gradually, instead of playing its accustomed role in metabolic activity, breaking down appropriate substances, building up others, the system goes awry. Some substances may no longer be broken down, but allowed to accumulate; others may be created in error, further disturbing the system's biochemistry.

Living systems have built-in redundancy, extra systems to take over when some fail; they can also carry out a considerable amount of repair work on deranged subsystems. Under some circumstances radiobiological damage is taken care of without ever becoming evident. But in others the initial disruptive effect precipitates consequences that multiply the disruption. Unfortunately, we still do not know exactly how the initial radiation damage triggers successive harmful consequences in living tissue. It is clear that a massive dose of radiation can simply overwhelm a living system, with so much primary damage that it is incapable of recovery. But much more insidious harm can be done even by a single alpha particle, beta particle, gamma ray or neutron, although it may take a very long time to develop: years or even decades. Its final manifestation may be totally unrecognizable as radiation damage, not only because of the time-lag but also because the pathological outcome may result from a very long train of cumulative biological consequences triggered by a random jolt to a minute but sensitive component. It can be appreciated, then, that the scientific study of radiobiology is challenging, frustrating, and open to widely differing interpretations of data.

The body whose standing is highest in the field of radiobiology as it affects decision-making is the International Commission on Radiological Protection (ICRP), founded in 1928, which is made up of leading radiobiologists from many different countries, and whose committees meet regularly to assess the current understanding of radiobiological phenomena. On the basis of such assessments the ICRP proposes standards for fields where radiation effects may arise. The ICRP's reports and recommendations form the basis of radiobiological standards in virtually all countries engaged in nuclear activities, although they are construed and applied differently in, for instance, the USA and the UK. ICRP Publication 9 is a concise presentation of the Commission's recommendations for radiation standards, and the reasoning upon which such standards are based. Other ICRP publications provide more detail, data and analysis.

Radiation effects can be described as acute or late depending on whether they show up within a matter of weeks of the radiation exposure, or only perhaps years afterwards. The effects can be further classified as 'somatic' and 'genetic'. A 'somatic' radiation effect shows up in the organism – perhaps the human body – which has been exposed. A 'genetic' effect shows up in the offspring or later descendants. Acute effects are fairly easy to identify as radiation injuries; late somatic effects may be much harder, and genetic effects may not be identifiable at all.

Acute radiation injury – hundreds of rads in a short time – causes damage to the tissues which form red blood cells; very high doses may also damage the stomach and intestines, and extreme doses the central nervous system. But smaller doses usually entail a longer sequence of biological consequences. Leukaemia may be induced five years or more after the exposure; other cancers may not show up until as much as twenty years after the exposure. Cataracts may form on the eye; there may be skin damage. Fertility may be impaired. At the level of virtual undetectibility is 'non-specific ageing' or 'radiation life-shortening', whose basis is very obscure. These are all somatic effects.

Even a single gamma ray can cause damage to a reproductive cell, to a gene or chromosome. If the damaged cell then participates in the formation of an offspring, the effect of the damage appears in the offspring – or, possibly, only in later generations. If the damage is sufficiently serious the offspring may not survive; if it does survive to reproduce, the so-called 'mutation' may slowly become a widespread feature of the descendant population.

We are continually subjected to ionizing radiation from natural sources: cosmic rays; uranium and thorium in the earth; and certain radioactive isotopes of substances in our bodies, particularly potassium-40. This 'background' radiation varies considerably from place to place on the earth, and according to height above sea level. It is usually of the order of 100 millirem (0.1 rem) per year. Since this natural radiation is inescapable, radiobiologists assume that we have learned to live with it biologically. This does not mean that it is harmless, merely that whatever harm it does allows us to exist without discernible ill-effects. Accordingly, the natural background is taken as a baseline for the setting of standards governing man-made radiation. The fundamental ICRP recommendations set down 'Maximum Permissible Doses' for those occupationally exposed to radiation, and 'Dose Limits' for members of the general public. Since radiation workers are aware of the possible hazards, and are expected to be supervised accordingly, the ICRP allows them more exposure; to allow a margin, the Dose Limits for the public are set ten times lower than the Maximum Permissible Doses for occupational exposure.

The ICRP identify the reproductive organs and the red bone-marrow as the most sensitive parts of the human body. The Maximum Permissible Dose for workers is set at 5 rem per year to either of these, and the Dose Limit for the public at 0.5 rem per year. Skin, bone and thyroid Maximum Permissible Doses are 30 rem per year, and Dose Limits 3 rem per year; hands and fore-arms 75 and 7.5, and other single organs 15 and 1.5. The Maximum Permissible Dose for the body as a whole is 5 rem per year for workers, and the Dose Limit for the public 0.5 rem per year.

At present, medical applications of radiation account for much the largest part of public exposure to man-made radiation. Medical and dental X-rays, and various forms of radiotherapy – notably radiation treatment for cancer – are instances in which a clear-cut benefit to the exposed individual is weighed against the statistical, varying, but small possibility of radiation injury. It is less easy to make a risk–benefit comparison in respect of other forms of man-made radiation. Fallout from nuclear weapons tests is a measurable contributor to present-day radiation exposure to the public, for benefits which must be regarded as debatable. This brings us squarely into the realm of controversy. What – if any – deleterious effects are caused by the low level of radiation from fallout? What – if any – deleterious effects are caused by the as yet far lower level of radiation resulting from running releases from civil nuclear installations? Are the radiation standards adequate to protect public health? Are they adequate to protect the health of workers in nuclear facilities?

We shall not here address the question of enforcement of standards, which is quite another matter. We shall rather examine two aspects of controversy over radiation standards, neither as yet resolved, to indicate points of contention.

The first aspect concerns the control of routine releases of radioactivity from nuclear facilities. There are two basic approaches to standard-setting for this purpose; one is employed in the UK, the other in the USA.

In the UK, when a proposal to release radioactivity is made, a survey determines where this radioactivity will go. Different radioisotopes follow different paths. When radioactivity is released into a waterway, some of it is deposited on the bed of the waterway, some washed ashore, some taken up by plants or by animals, and so on. A radioisotope which is initially dilute in the original discharge may be concentrated by organisms which absorb it. All these possibilities must be identified and assessed. For the planned discharge a 'critical group' of ultimate human consumers is distinguished, those whose consequent exposure is likely to be greatest, for radioactivity following the 'critical path'. If the critical group's exposure is to be kept below the ICRP Dose Limit, this implies a restriction on the original discharge, called a 'derived working limit'. The maximum permitted rate of radioactive discharge is set accordingly, and regular monitoring ensures that the limit remains within the ICRP criteria. At the time of writing there has been no scientific challenge to this procedure, which has been in effect ever since the early days of nuclear activity in Britain.

In the USA a different approach is taken. Standards are set on a nation-wide basis. For any particular radioisotope there is laid down a Maximum Permissible Concentration in air (MPC_a) or water (MPC_w). No effluent discharged from a nuclear facility must exceed this concentration at the site boundary. Clearly, it is easier for some facilities than for others to meet these criteria; some nuclear installations discharge effluents whose concentrations are a significant fraction of the maximum permissible, while others discharge effluents far below the permissible concentrations. Standards set on this basis became the target for detailed critical analysis by Dr John Gofman and Dr Arthur Tamplin. The National Committee on Radiological Protection (NCRP), a private body which now has government approval to set standards, advises the AEC (now NRC) which has the statutory responsibility for applying the standards. The NCRP and AEC were plunged into heated argument with Drs Gofman and Tamplin, both expert in the field of radiation biology, who proposed the concept of a 'doubling dose': the dose of radiation which might be expected to double the incidence of any pathological effect. The arguments involved are subtle, and the statistical data open to differing interpretations; but the work of Drs Gofman and Tamplin led them to assert that current AEC effluent standards would lead to something like 24 000 additional cases of cancer in the USA per year. The AEC challenged the validity of the 'doubling dose'; to suggest that the whole population would receive anything remotely approaching the maximum permissible dose was, they asserted, nonsense. To this the critics' response was – if so, why not lower the standards? In due course, for certain radioisotopes, the standards were indeed lowered. But the debate continues.

More recently, Dr Tamplin and Dr Tom Cochran published a report suggesting that inhaling microscopic dust particles of plutonium might have much more serious effects than current standards indicate. In their view the very highly concentrated short-range alpha emission from such a 'hot particle' delivers a very high radiation dose to a tiny local volume of lung tissue; the effect of such intense local radiation cannot, they suggest, be adequately anticipated on the basis that it is averaged over the whole mass of the lung, as present standards dictate. Drs Tamplin and Cochran insist that their analysis implies a drastic reduction in the permissible concentration of plutonium oxide in air – reduction by a factor of 115 000. Again, their analysis is vigorously disputed, both in the USA by the AEC, and in the UK by the National Radiological Protection Board. But again the debate continues. (See the Bibliography, p. 294, for further material.)

On a broader front, criteria for man-made radiation must look forward to the quantities of radioactivity which planned nuclear developments entail. Studies of the 'dose commitment' of radioactivity already released indicate that margins may become very much narrower before the end of this century. For the moment the most pressing requirement is better data, more careful compilation of records of exposure and medical histories of people known to have encountered man-made radiation of various kinds. Radiation experts point out repeatedly that it is not satisfactory to extrapolate uncritically from experiments on animals to forecast effects on human beings. But since few people would countenance planned experimentation on human subjects it ought to be a basic tenet of radiation medicine to collect and compile detailed records of effects where they occur or may occur. To take an obvious instance: the USA has established since 1968 a 'Transuranium Registry' to keep track of employees who have encountered plutonium and other actinides in the course of their employment, and to follow their subsequent medical histories. In Britain such record-keeping has been undertaken only in 1975. Until better evidence is available radiobiology will remain a hotbed of profoundly frustrating controversy.

Appendix C

Bibliography: A Nuclear Bookshelf

For those wishing to learn more about nuclear reactors and their world, the following sources are variously useful. Many of them will in turn indicate others.

The US Atomic Energy Commission (AEC) during its twenty-eight years published an enormous range of material, from the unintelligible to the trivial, with a vast amount of essential information in between. Its successors, the Energy Research and Development Administration (ERDA) and the Nuclear Regulatory Commission (NRC) are now taking on the task. The AEC's series of popular booklets *Understanding the Atom* are free and informative, provided you read scrupulously between the lines. The sweetness and light is like an overdose of whipped cream. At a more profound level are the AEC documents bearing the WASH prefix. Curiously enough the AEC never published a catalogue of these, and the numbers bear little relation even to chronological order. But many of the sources mentioned below cite WASH documents by number and title in many areas of importance. Other AEC/ERDA/NRC publications of value are the weekly compilations of News Releases, the monthly series *Current Events – Power Reactors* (offshoot of the series *Reactor Operating Experience*), *Safety Related Occurrences*, *Regulatory Guides*, *ad hoc* reports on important developments – the list could be expanded for pages. The AEC's successors maintain a Public Document Room at the offices at 1717 H Street, NW, Washington DC, at which all public AEC/ERDA/NRC documents are available for inspection; photocopying facilities are also provided. Write to ERDA at Washington DC 20545, USA, and the NRC at Washington DC 20555, USA. These addresses reach all internal departments. Inquirers can often obtain free a single copy of a new ERDA or NRC publication; the News Releases give details. All AEC/ERDA/NRC documents are available for purchase from the National Technical Information Service of the US Department of Commerce, 5285 Port Royal Road, Springfield, Va. 22151, USA.

The International Atomic Energy Agency (IAEA) also publishes a wide range of material. For hard data the shelves filled with the *Proceedings* of the four Geneva Conferences on the Peaceful Uses of Atomic Energy are invaluable, albeit technical. The IAEA Directory is a – thus far – eight-volume compilation of data on the world's reactors; the paperback *Power and Research Reactors in Member States* is a handy compendium. The IAEA also publishes proceedings of the many specialized conferences it organizes, and a free and useful bi-monthly Bulletin. IAEA publications are available from its Division of Publications, or from government publications outlets in other countries – in Britain, for instance, from Her Majesty's Stationery Office, PO Box 569, London SE1 9NH.

The OECD Nuclear Energy Agency publishes conference proceedings and reports, and an annual *Activity Report*, available from the Nuclear Energy Agency's Paris office or from government outlets in other countries. Typical recent publications of value include an excellent concise survey of *Radioactive Waste Management in Western Europe*, a comprehensive introduction to the subject, and *Management of Radioactive Wastes from Fuel Reprocessing*, a conference report which comprises in one – admittedly expensive – volume a complete picture of the state of the art world-wide, a classic source of detailed information.

Two other official bodies whose publications should be noted are the United Nations Scientific Committee on the Effects of Atomic Radiation, who publish a major survey on *Ionizing Radiation*; and the International Commission on Radiological Protection, whose Publication 9 lays down the bases for most national restrictions on radioactivity, and whose other publications are the standard international references on radiobiology.

The United Kingdom Atomic Energy Authority publishes, among many other items, the concise directory *Reactors UK* giving technical details on all British reactors. Their monthly bulletin *Atom* is also informative. Other national nuclear authorities should also be noted as sources for appropriate information. In Britain the National Radiological Protection Board and, on occasion, the Inspectorate of Nuclear Installations, publish material of interest, available through HMSO.

Other official sources include the various US Congressional hearings, far too numerous to list, and, in Britain, certain of the sittings of the Parliamentary Select Committee on Science and

Technology. Background papers for the US Congress – again too numerous to list – are also worthy of note; they are obtainable from the Superintendent of Documents, US Government Printing Office, Washington D C 20402. Ask whether the particular topic in which you are interested has been covered. The Library of Congress in Washington is of course another invaluable repository of information.

Industrial and commercial associations of nuclear interests, like the Atomic Industrial Forum in the USA and the British Nuclear Forum in Britain, offer various information documents and periodicals. On a more academic level organizations like the American Nuclear Society, the British Nuclear Energy Society and the Canadian Nuclear Association publish conference proceedings and reports of value.

The most concentrated compilation of basic nuclear physics and engineering is undoubtedly the *Source Book on Atomic Energy* by Samuel Glasstone (third edition, Van Nostrand Reinhold, 1968): a one-volume encyclopaedia, clear without avoiding technicalities. It does, however, omit most of the more troubling aspects of the story. Much more technical is *Nuclear Reactor Engineering* (Van Nostrand Reinhold, 1969) by Samuel Glasstone and Alexander Sesonske, a standard work on the subject, loaded with valuable information and data but very much a specialized textbook.

Britain's Central Electricity Generating Board offers, in *Modern Power Station Practice*, Volume 8: *Nuclear Power Generation* (Pergamon, 1971), a superbly detailed and thorough discussion of how to design, build and operate your very own reactors. Not for the complete layman, but accessible; and honest about problems, at least on the engineering side.

In late 1974 there appeared *Independence and Deterrence*, by Margaret Gowing, assisted by Lorna Arnold (Macmillan, London, 1974), the official history of British nuclear development from 1945 to 1952. This is by any criterion an extraordinary book, massively detailed yet readable, even gripping. Volume 1 deals with *Policy Making*, Volume 2 with *Policy Execution*. Together they describe how Britain acquired nuclear weapons virtually in secret, and how the scientists and engineers created a vast industry with almost no guidance or direction by the politicians. Despite its size and cost this book is essential reading for anyone anxious to unravel the way nuclear energy has developed. Professor Gowing's earlier book, *Britain and Atomic Energy, 1939–45* (Macmillan, London, 1964) the first phase of the official history, is likewise invaluable. It must be hoped that the forthrightness of her commentary does not lead to official 'deterrence' of a proposed third phase of the official history.

The official histories of US nuclear development began with *A General Account of the Development of Methods of Using Atomic Energy for Military Purposes under the Auspices of the United States Government 1940–1945* by Professor H. D. Smyth, published in August 1945 at the behest of Major-General Leslie Groves, Director of the Manhattan Project. For obvious reasons the report has since become known simply as the Smyth report. Later editions were published by the Princeton University Press. The Smyth report gives a fascinating account of the physics and – to some extent – the politics of the first nuclear programme, fascinating especially in the light of subsequent knowledge. The official histories of the US Atomic Energy Commission and its forerunners are *The New World 1939–1946* by R. G. Hewlett and O. E. Anderson (Pennsylvania State UP, 1962), and *Atomic Shield 1947–1952* by R. G. Hewlett and F. Duncan, (Pennsylvania State UP, 1969). More recently, C. Allardice and E. Trapnell, long-time AEC commentators, published *The Atomic Energy Commission* (Praeger, 1974). The various US official histories, unlike those of Professor Gowing about the UK, tend to gloss over controversies and present strictly 'official' interpretations. Differing viewpoints are available in some of the following sources.

One of the first – and still one of the best – historical accounts of the early years of nuclear development is *Brighter than a Thousand Suns* by Robert Jungk (Penguin, 1970). Subtitled *A Personal History of the Atomic Scientists* it recounts a vividly readable narrative of the men who made the A-bomb and the H-bomb, in particular Robert Oppenheimer and Edward Teller. A more recent retelling, focusing on many of the same people, this time by an American writer, is *Lawrence and Oppenheimer* by Nuel Pharr Davis (Jonathan Cape, 1969) – again vivid and readable, by no means a recap of Jungk, offering different but equally provocative insights into policies and personalities, and the context in which the bombs were created. Yet a third history is *Men Who Play God* by Norman Moss (Penguin, 1970), which picks up the story somewhat later and carries it through most of the 1960s, detailing not only the US and British work on thermo-nuclear bombs but including material about the opponents of these developments, and about the theoreticians like Hermann Kahn who devised their intellectual context. Elizabeth Young, in *A Farewell to Arms Control?* (Penguin, 1972) describes the government and diplomatic context of the nuclear arms race, including concise summaries of the policy issues and their development in the participant countries. Ralph Lapp's classic *The Voyage of the Lucky Dragon* (Penguin, 1958) tells the story of the luckless Japanese fishermen caught by the fallout from the Castle Bravo H-bomb test; this was probably the first important popular book to challenge the AEC's policies and their execution.

Lapp's classic came out about the time of the first Pugwash conference. The 24th took place in 1974. The published proceedings of all the conferences tell an inside story of the on-going struggle to devise credible international policies for control of nuclear energy in all its manifestations. Pugwash also publishes a quarterly newsletter. The Stockholm International Peace Research Institute, established in 1966 by the Swedish govern-ment to commemorate 150 years of peace in Sweden, is now a world authority on *World Armaments and Disarmament*, the title of its annual *Year Book*, a massive study of military policy, technology and financing, and of its national and international context. The institute also publishes monographs on particular subjects, like *Nuclear Proliferation Problems* (1974). Institute publications are available from Almqvist & Wiksell, PO Box 62, S-101 20 Stockholm 1, Sweden, and from publishers Paul Elek, Humanities Press and the MIT Press.

A more laconic commentary on nuclear military developments is provided by the International Institute for Strategic Studies in *The Military Balance*, published each autumn, and *Strategic Survey*, published each spring. Their publications, which include many monographs on related subjects, are available from 18 Adam Street, London WC2N 6AL.

Two of the best introductory books on radiation are *Radiation: What it is and How it Affects You* by Jack Schubert and Ralph Lapp (Heinemann, 1957), and *Atomic Radiation and Life* by Peter Alexander (Penguin, 1957); unfortunately both of these are out of print, but worth seeking out. More scientifically detailed – and of course more up-to-date – is *Biological Effects of Radiation* by J. E. Coggle and G. R. Noakes (second edition, Wykeham

Publications, 1972), a business-like and hard-headed presentation of the facts as known and their interpretation, and the questions still unanswered.

Radioactive Contamination (Harcourt Brace Johanovich, New York, 1975) by Virginia Brodine, consulting editor of *Environment*, puts the subject of radiobiology in its technical, political and economic context, as an issue not merely of academic interest but of public policy. As an introduction to the subject it is intelligible, informative and matter-of-fact – highly recommended.

Other books provide more outspoken views on radiation. *Poisoned Power: The Case Against Nuclear Power Plants* (Chatto & Windus, 1973) is an unattractive title, but most people who know how the A E C has treated the book's distinguished authors, Drs John Gofman and Arthur Tamplin, will excuse the title. The book is an angry – but accurate – challenge to the A E C from within its most qualified ranks; it takes issue with many aspects of A E C policy, in particular radiation standards, and has already been instrumental in tightening those standards. Anyone wishing to comprehend the dimensions of nuclear dissent must be familiar with the critiques of Gofman and Tamplin. A yet more contro-versial view is put forward by Dr Ernest Sternglass in *Low Level Radiation* (Ballantine, New York, 1973). His thesis on the hazards from even slight additions to environmental radiation has been variously attacked and defended from all sides; if his thesis is even partially sound we are already in serious trouble.

Undoubtedly the most authoritative work on reactor problems is *The Technology of Nuclear Reactor Safety* (M I T Press, 1965 and 1973), edited by Drs Theos Thompson and J. G. Beckerley of the A E C. It covers *Reactor Physics and Control* in Volume 1, and *Nuclear Materials and Engineering* in Volume 2. Although frighteningly expensive they are the most comprehensive, indeed exhaustive compilation available, with a wealth of historical information both technically detailed and scrupulously complete. Thompson and Beckerley pulled no punches; when poor design, short-sightedness, corner-cutting or carelessness led to trouble they were uncompromising. The first popular book on reactor problems was probably Sheldon Novick's *The Careless Atom* (Delta, New York, 1970). It is still one of the best, readable, purposeful and business-like. Novick's senior colleague Barry Commoner was early in the field of nuclear controversy with sections of *Science and Survival* (first edition, Viking, 1965;

revised edition, Ballantine, 1971) and later with sections of *The Closing Circle* (Jonathan Cape, 1972) - the latter a bit too ideologically oriented for my own taste. Still less satisfactory is *Perils of the Peaceful Atom* (Gollancz, 1970) by Richard Curtis and Elizabeth Hogan, unrelentingly shrill and strident and with tendentious lapses in accuracy. *The Gentle Killers*, by Ralph Graueb (Abelard Schuman, 1974) is characteristic of another genre of nuclear criticism, juxtaposing nuclear hazards with enthusiastic espousal of dietary and other reforms, coming out cranky in all directions. An Oregon reporter named Gene Bryerton compiled a useful, relatively calm-voiced series of newspaper articles into *Nuclear Dilemma* (Ballantine, New York, 1970), giving the proponents as well as the opponents of nuclear power stations space to state their opinions. One of the most unexpected contributions to the reactor dossier comes from George L. Weil, who as a young physicist started up Fermi's first reactor in 1942. Weil's subsequent disillusion with reactors led him to publish privately a concise, wrathful small book entitled *Nuclear Energy: Promises, Promises*, delineating the main problems faced - or ignored -'by the US nuclear industry. It is obtainable from him at 1730 M Street, NW, Washington DC 20036, price $2.00.

Professor David Inglis also has a long career in the nuclear field to his credit, and his *Nuclear Energy: Its Physics and Its Social Challenge* (Addison-Wesley, 1973) covers the broadest imaginable spectrum. It does so, however, unevenly, interweaving slices of elementary physics textbook with snatches of history, social, political and ethical comment which somehow - to me at any rate - fail to carry total conviction. A more down-to-earth summary of the essentials of nuclear power generation is given in *Energy* (Sierra Club, New York, 1973) by John Holdren and Philip Herrera; the first half of their book presents an overall introduction to the technologies of energy supply, including nuclear, and the second half examines a series of case-histories of controversies arising from the said technology, including Calvert Cliffs, Bodega Head and Monticello.

The year 1972 saw the challenge to the AEC escalate into a barrage. *The Great American Bomb Machine*, by Roger Rapoport (Dutton, New York, 1971) is a brilliant dissection of the most unpleasant innards of the AEC's nuclear weapons activities. *The Atomic Establishment* by H. Peter Metzger (Simon & Schuster,

New York, 1972) is a scathing historical critique of the AEC in action, and of its cosy relationship with the Congressional Joint Committee on Atomic Energy, the watchdog that 'did nothing in the night-time'. *The Nuclear-Power Rebellion: Citizens vs. the Atomic Industrial Establishment* by Richard Lewis (Viking, New York, 1972) is an inside look at the many battlefronts upon which the US civil nuclear industry and its critics have clashed. The author, as Editor of the *Bulletin of the Atomic Scientists*, had a unique opportunity to observe the confrontations and indeed to further them. *Citizen Groups and the Nuclear Power Controversy* by Stephen Ebbin and Raphael Kasper (MIT Press, Cambridge, Mass., 1974) analyses how public participation in US nuclear licensing procedures has worked – or failed to work – in general, and taking a closer look at three hearings including the one on emergency core cooling systems. There do not seem to be any equivalent studies in other countries – probably because public participation has played an even smaller role elsewhere than it has in the USA.

The Union of Concerned Scientists (1208 Massachusetts Avenue, Cambridge, Mass. 02138) has published a series of pungent and detailed critiques, particularly of the emergency core cooling system issue and of the nuclear fuel cycle. *Preventing Nuclear Theft: Guidelines for Industry and Government* edited by Robert Leachman and Philip Althoff (Praeger, 1972), the proceedings of the 1971 symposium at Kansas State University, was one of the first books to go into detail about the problem of keeping track of fissile materials. *International Safeguards and Nuclear Industry*, edited by Mason Willrich (Johns Hopkins University Press, Baltimore, 1973), covered many of the same categories of problem, as they manifested themselves from the Non-Proliferation Treaty all the way to the lunatic fringe. Mason Willrich also wrote *The Global Politics of Nuclear Energy* (Praeger, New York, 1971). But probably the definitive book on the subject is his collaboration with Ted Taylor for the Ford Foundation Energy Policy Project, *Nuclear Theft: Risks and Safeguards* (Ballinger, Cambridge, Mass., 1974), a masterly, low-key presentation of the whole gamut of considerations: what fissile material will be available to whom under what circumstances, what might be done with it by whom, and what can be done to keep it out of the wrong hands. Unfortunately this final section is much the least persuasive. John McPhee wrote a profile of Taylor for the *New*

Yorker; the resulting book, *The Curve of Binding Energy* (Dutton, New York, 1974) is a translation of the Energy Policy Project book into human terms, grimly convincing and offering little compensatory consolation. In McPhee's view the next nuclear explosion in a city seems to be only a matter of time – and not a very long time.

A sizeable sector of the world's nuclear industry is now working on the assumption that future energy supplies will derive in large measure from fast breeder reactors. In the USA the validity of the AEC's position on this issue is challenged in *The Liquid Metal Fast Breeder Reactor: An Environmental and Economic Critique* by Tom Cochran (Resources for the Future, Baltimore, 1974), in which the economic justification of this particular nuclear option is subjected to searching cross-examination and found wanting. Since a nuclear power system based on fast breeder reactors would involve plutonium by the tonne, Dr Cochran also expresses some serious reservations akin to those of Willrich and Taylor. In 1974 Dr Cochran was also co-author of a critique of the radiation standards for 'hot particles' of plutonium and other actinides, on behalf of the Natural Resources Defense Council. This report drew an AEC rebuttal, WASH-1320, to which Cochran and Tamplin in turn replied. Their 'hot particle' analysis also formed part of the Council's critique of the AEC draft environmental impact statement on the fast breeder programme, WASH-1535. The draft statement attracted much unfavourable comment, including that of the US Environmental Protection Agency, and has not at the time of writing reappeared in final form.

In the UK, the considerations affecting energy policy, including the role of nuclear energy, were described in historical detail with wit and insight by Michael Posner in *Fuel Policy: A Study in Applied Economics* (Macmillan, London, 1973). A good many of his assumptions have since, however, become less convincing; a second, revised, edition would certainly be welcome. Lord Sherfield edited *Economic and Social Consequences of Nuclear Energy* (Oxford University Press, 1972), a collection of essays presenting the views of, among others, Sir Stanley Brown, chairman of the Central Electricity Generating Board during some of its major nuclear expansion. It is instructive to examine the considerations which then influenced British policy *vis-à-vis* nuclear energy. *Nuclear Power* by W. G. Jensen (Foulis, 1969), further helps to place the role of nuclear energy in its economic and political context, giving valuable background to the European situation in particular. Jensen is, however, neither physicist nor biologist, and the economics is on a traditional and uncomfortably narrow foundation.

I must also – first declaring my interest, in that the authors are valued colleagues of mine – recommend *World Energy Strategies* by Amory Lovins (Friends of the Earth, London, 1974; Ballinger, Cambridge, Mass., 1975), *Nuclear Power: Technical Bases for Ethical Concern* also by Amory Lovins (Friends of the Earth, London, 1974), and *Dynamic Energy Analysis and Nuclear Power* by John Price (Friends of the Earth, London, 1974); the latter two titles are included in *Non-Nuclear Futures*, edited by Amory Lovins (Ballinger, Cambridge, Mass., 1975). Related papers include *Energy Inputs and Outputs for Nuclear Power Stations* by Peter Chapman and Nigel Mortimer (Energy Research Group, Open University, research report 005, revised December 1974) and *Nuclear Energy Balances in a World with Ceilings* by Gerald Leach (International Institute for Environment and Development, London, December 1974). Dr Chapman has more recently produced a witty and thought-provoking analysis of 'energy options for Britain' called *Fuel's Paradise* (Penguin, 1975). In it he investigates possible future patterns of energy supply and demand in their social and economic context, pointing out the various constraints and stumbling-blocks which lie before us according to which pathway we choose.

In a field changing as swiftly as that of nuclear energy, periodical publications also constitute essential sources of information. The monthly *Bulletin of the Atomic Scientists*, 1020–24 E 58th Street, Chicago, Ill. 60637, USA was founded in 1945; in the thirty years of its existence it has been a forum for consistently far-sighted and thoughtful discussion of the implications of nuclear energy for the world. *Environment*, 560 Trinity Avenue, St Louis, Mo. 63130, USA, monthly, was founded in 1958 as *Nuclear Information*; despite the wider concerns its title suggests it continues to pay close and critical attention to the nuclear arena. The weekly *New Scientist*, 128 Long Acre, London WC2E 9OH, UK, is a lively and readable magazine covering all aspects of science and society, with regular dispatches about energy technology and policy, including nuclear energy. *Nature* (4 Little Essex Street, London EC1, UK) and *Science* (1515

Massachusetts Avenue N W, Washington D C 20005, U S A), weekly, include, as well as original scientific papers and notes, regular news items which include major coverage of the nuclear field. The quarterly journal *Energy Policy*, 32 High Street, Guildford, Surrey G U I 3E W, U K, although expensive, contains valuable in-depth articles by leading authorities on energy in general and nuclear energy in particular. Then there are the four periodicals which should be known to all those concerned with nuclear energy. *Nucleonics Week* is an international weekly newsletter published by McGraw-Hill (1221 Avenue of the Americas, New York, NY 10021), devoid of advertising and often sufficiently outspoken to bring aggrieved looks from the industry. But its subscriptions price is more than $350 per year . . . check your library. *Nuclear News*, published monthly by the American Nuclear Society (244 East Ogden Avenue, Hinsdale, Illinois 60521, U S A), covers the whole field of nuclear energy in depth and – considering its origins – gives balanced commentary on issues and controversies. *Nuclear Engineering International* is a glossy monthly offering issue-length features on new nuclear plants, or on particular aspects of the technology; its annual output includes a March issue devoted to an exhaustive international list of firms offering nuclear materials and services, and an April issue with a detailed directory of the world's reactors. It also publishes fold-out wall-charts of nuclear plants and other information. My own favourite of these three is the *Weekly Energy Report*, published independently at 1239 National Press Building, Washington D C 20045, by Llewellyn King. The *Weekly Energy Report* is a newsletter free of advertising, and, once again, accordingly very expensive – $325 per year. But with that proviso it is nonetheless an authoritative source which manages to maintain its credibility and its credit across the whole panorama of energy policy, including nuclear policy, no mean accomplishment. It ought to be in all major libraries.

Appendix D

Nuclear Organizations Pro and Con

International Atomic Energy Agency, Kaertner Ring 11, PO Box
590, A-1011 Wien, Austria.

OECD Nuclear Energy Agency, 38 Boulevard Suchet, 75016
Paris, France.

United States Nuclear Regulatory Commission (Washington
DC 20545) and the Energy Research and Development
Administration (Washington DC 20555) now embody the
former US Atomic Energy Commission.

United Kingdom Atomic Energy Authority, 11 Charles II Street,
London SW1, UK.

Commissariat à l'Energie Atomique, BP 510, 75752 Paris Cedex,
France.

Atomic Energy of Canada Ltd, 275 Slater Street, Ottawa,
Canada K1A 0S4.

(For other national nuclear organizations, government and industry, try embassies for details.)

United Nations Scientific Committee on the Effects of Atomic
Radiation, United Nations, New York, USA.

International Commission on Radiological Protection, Clifton
Avenue, Block D, Belmont, Surrey, UK.

National Radiological Protection Board, Harwell, Didcot,
Oxfordshire, OX11 ORQ, UK.

Atomic Industrial Forum Inc., 1747 Pennsylvania Avenue NW,
Washington DC, 20006, USA.

British Nuclear Forum, Leicester House, 8 Leicester St., London
WC2, UK.

American Nuclear Society, 244 East Ogden Avenue, Hinsdale,
Illinois 60521, USA.

British Nuclear Energy Society, 1 Great George Street, London
SW1, UK.

Canadian Nuclear Association, Suite 1120, 65 Queen Street W,
Toronto, Ontario M5H 2M5, Canada.

Stockholm International Peace Research Institute, Sveavaegen
166 S-113 46 Stockholm, Sweden.

Pugwash, 9 Great Russell Mansions, 60 Great Russell Street, London WC1, UK.

Friends of the Earth Ltd, 9 Poland Street, London WIV 3DG, UK.

Friends of the Earth Inc., 529 Commercial Street, San Francisco, Cal. 94111, USA.

Les Amis de la Terre, 16 rue de l'Université, 75007 Paris, France.

Jordens Vaenner, Box 9062, Heleneborgsgatan 15B, S-102 71 Stockholm, Sweden.

Vereniging Milieudefensie, Herengracht 109, Amsterdam, the Netherlands.

Friends of the Earth (Ireland), 1 Slane Road, Crumlin, Dublin, Ireland.

Freunde der Erde, 2 Hamburg 1, Postbox 100221, West Germany.

(Other Friends of the Earth national organizations can be reached via any of the above.)

The Conservation Society, 12 London Street, Chertsey, Surrey, KT16 8AA, UK.

The Sierra Club, 1050 Mills Tower, San Francisco, Cal., USA.

Scientists' Institute for Public Information, 30 East 68th Street, New York, NY 10021, USA.

Natural Resources Defense Council Inc., 917 15th Street NW, Washington DC 20005, USA.

Union of Concerned Scientists, 1208 Massachusetts Avenue, Cambridge, Mass. 02138, USA.

Businessmen for the Public Interest, Suite 1001, 109 North Dearborn, Chicago, Illinois 60602, USA.

Center for the Study of Responsive Law (Ralph Nader), PO Box 19367, Washington DC, 20036, USA.

Index

See also Appendix A 'Nuclear Jargon', Appendix C 'Bibliography', and Appendix D: 'Nuclear Organizations'. Names – such as US, UK, USSR, Canada, uranium, plutonium, AEC, AEA, PWR and BWR – which recur repeatedly throughout the text are not listed separately but only under other specific references.

'Able' test, 122
Abrahamson, D., 151
advanced gas-cooled reactors (AGRs), 55ff
Advisory Committee on Reactor Safeguards, 169–70, 182
Agesta, 184, 243
Alamogordo, 32
Alaska, 153, 155
Aldermaston, 143–4
Allied Gulf, 208
Almelo, 94, 227
Amchitka, 153
Anderson, C., 170
Anjain, L., 134
APS-1, 49
Aquafluor, 208–9
Argentina, 245, 248, 253, 255, 257
Arms Control and Disarmament Agency, 249, 265
ASEA, 184–5
Asse, 114
Atlantic–Pacific Inter-Oceanic Canal Study Commission, 155
Atlantic Richfield, 110–11, 215
Atomic Energy Research Establishment, 130
Atomic Industrial Forum, 216
Atomic Power Constructions, 183
Atomic Weapons Research Establishment, 143–4
'Atoms for Peace', 129, 173
Atucha, 248
Australia, 87, 128, 153, 154, 246, 253, 257, 267
Austria, 257
Avoine, 55
AVR, 59

Babcock & Wilcox, 197
Baby Tooth Survey, 146
'Baker' test, 122

Baltimore Gas & Electric, 192
'Baneberry', 153
Bangladesh, 257
Bank, A., 252
Barnwell, 209
Bartlett, A., 236
Battelle, 200
Becquerel, H., 117
Belgium, 109, 244, 246, 253, 257
Bergkrankheit, 120
Berkeley, 53, 162, 166, 183, 222
Bethe, H., 168, 216
Big Rock Point, 167
Bikini, 122–3, 124, 133
biogeneration, 267
Black, H., 179
BN-350, 77, 206
BN-600, 207
Bodega, 179
boiling water reactors (BWRs), 66ff
BONUS, 167
BORAX, 175
Bradwell, 53, 162, 166, 183, 222
Brazil, 245, 253, 257
Brennan, 178
British Nuclear Design & Construction, 184
British Nuclear Fuels Ltd, 207, 227–8
Brookhaven National Laboratory, 170, 200, 236
Browns Ferry fire, 214–15
Bruce, 70, 213
Bulgaria, 257
Bupp, I., 231–2
Businessmen for the Public Interest, 209, 225
Byrnes, J., 175–6

Calder Hall, 50ff, 61, 130, 161, 162, 164ff, 222
Calvert Cliffs, 192–3

Campaign for Nuclear Disarmament, 143-4
CANDU reactors, 70ff, 75, 94, 97, 158, 202, 213-14, 255
'Cannikin', 153
Capenhurst, 91, 94, 227
Carlsbad Caverns, 114
'Castle Bravo', 129, 133
CENTEC, 227
Central Electricity Generating Board, 53, 54, 162, 166, 183-4, 200ff, 219, 223ff, 266
Chad, 246
chain reaction, 30ff
Chalk River, 70, 158, 160
Chapelcross, 50, 165, 166, 222
Chariot, Project, 155
Chicago Pile No. 1, 54
Chile, 245, 257
China, 18, 91, 109, 147, 154, 244, 245, 247
Chinon, 55, 161
CIRUS, 244, 254
Clark, H., 132-3
Clinch River, 77, 205
Cochran, T., 285
Cockcroft, J., 163
Cole, L., 146
Colorado, 59, 90, 156, 240
Combustion Engineering, 197, 200
Comey, D., 209, 225
Commissariat d'Energie Atomique, 161, 259
Committee for Nuclear Information, 143, 155
Commoner, B., 146
Commonwealth Edison, 172, 178
Connecticut Light & Power, 178
Consolidated Edison, 172, 178
Consolidated National Intervenors, 197
Consumers Power, 178, 200
Cooperative Power Reactor Demonstration Program, 129, 167
Cottrell, A., 201
Crossroads, 123
Cuba, 147, 245, 248, 256
Curie, 28, 119
CVTR, 167
Czechoslovakia, 257

Daghlian, H., 121
Denmark, 257
Derian, 231-2
Detroit Edison, 77, 168ff, 182
Dimona, 243
Disarmament Conference, 243
Dose Limit, 282ff
Douglas, W., 179
Dounreay, 77, 166, 249, 253

Dounreay Fast Reactor, 77, 79, 84
Dow, 238
Dragon, 59, 61, 173
Dresden, 167, 172, 178, 189ff
Dubos, R., 146
Dungeness, 183-4, 223-4
Duquesne Power & Light, 166
du Pont, 215

East Germany, 257
Eastman Kodak, 132
Eazor Express Corporation, 236
EBR-1, 76, 77, 167-8, 175
EBR-2, 77, 203
Edlow, S., 249
Egypt, 246, 253, 255, 257
Einstein, A., 139
Eisenhower, D., 129, 146, 173
EL-1, 161
Electricité de France, 162
Elk River, 167
emergency core cooling systems, 65, 69, 194ff
Engineering Research Institute, 171
Eniwetok, 128
enrichment, 91ff
Enrico Fermi-1, 77, 167ff, 178-9, accident, 180ff, 188, 204
Environmental Protection Agency, 192, 205
Euratom, 174
Eurochemic, 173
European Coal and Steel Community, 174
European Economic Community, 174, 259
European Nuclear Energy Agency, 173
Exxon, 215

fallout, 132ff, 153
fast breeder reactors (FBRs), 75ff
Fast Flux Test Facility, 77, 204
Federation of Atomic Scientists, 122
Fermi, E., 45
Fessenheim, 189
Finland, 257
Fontenay-aux-Roses, 161
Ford, D., 197
Ford, G., 231
Fort St Vrain, 59
Franck, J., 121
Friends of the Earth, 200, 201, 212
fuel fabrication, 97-8
fuel reprocessing, 100ff
Fukuryu Maru, 133, 135, 212
fusion, 126

G-1, 161
gas centrifuge, 94

gas-stimulation, 156
'Gasbuggy', 156
gaseous diffusion, 91ff
General Accounting Office, 157, 250
General Atomic, 59
General Electric (US), 151, 183, 196-7, 208, 215, 223, 227
General Electric Company (UK), 201
Geneva Conference on the Peaceful Uses of Atomic Energy, 170
Gentilly-1, 73
geothermal energy, 267
Gillette, R., 198
Girdler–Sulphide process, 95
Glace Bay, 213
Gleason, E., 235-6
Gofman, J., 148, 150-51, 152, 284-5
Goodman, L., 170
Grand Junction, 90, 156
Greece, 257
Greenpeace, 153, 154
Grosswelzheim, 174
Gulf, 215

Haddam Neck, 178
Haiti, 246
Halden, 173
Hallam, 167
Hanford, 48, 66, 77, 109ff, 149, 175, 237, 241
Hartlepool, 184
Harwell, 130, 218
Hawkins, A., 200
Hayes, H., 149
Heat Transfer Reactor Experiment, 175
Heath, E., 201
heavy water production, 95ff
Heysham, 184
high temperature gas-cooled reactors (HTGRs), 57ff
Hinkley Point, 182, 184
Hinton, C., 201
Hiroshima, 23, 32, 91, 120, 121, 122, 124, 127, 135
Holifield, C., 170, 231
Hosmer, C., 227, 230
'hot particle', 285
Humboldt Bay, 167
Hungary, 128, 257
Hunterston, 54, 183, 184
hydrogen bomb, 126, 128, 135, 144

Idaho Nuclear Corporation, 196
Iklé, F., 265
India, 96, 109, 135, 244ff, 253ff, 257
Indian Point, 167, 172, 200
Indonesia, 246
Institute of Nuclear Materials Management, 249

Institution of Professional Civil Servants, 201
International Atomic Energy Agency, 173, 199, 242, 254, 259
International Commission on Radiological Protection, 90, 108, 281ff
Iran, 257
Israel, 243, 245, 253, 255ff
Italy, 50, 175, 189, 253, 257

Jamaica, 257
James Bay, 94
Japan, 16, 50, 120, 121, 153, 174, 175, 189, 210, 246, 253, 257, 261, 267
Joachimsthal, 90, 119-20
Joint Committee on Atomic Energy, 124, 142, 145, 152, 170, 227, 231
Jordens Vänner, 212
Jülich, 59

Kalckar, H., 146
Kansas, 113
Kendall, H., 197
Kennedy, J., 147
Khrushchev, N., 146, 147
Knapp, H., 145, 147
Korea, 245, 246, 257
Kraznoyarov, N., 206
Kurchatov, I., 128
Kwajalein, 134

La Crosse, 167
Langham Committee, 148
Lapp, R., 139, 143, 216
laser enrichment, 95
Latina, 175
Lawrence Livermore Laboratory, 148, 149, 150, 200
Legg, R., 175-6
Lenin, 210
Lesotho, 246
Lewis, E., 142
Lilienthal, D., 180
Lop Nor, 244
Los Alamos, 121, 122, 125, 140
Loss-of-Fluid Test, 196
Lucens accident, 185ff
Lucky Dragon, 135, 143
luminizers, 119
Lyons, 113

McKinley, R., 175-6
McMahon Act, 124-5, 127
Madagascar, 246
Magnox reactors, 49ff
Malibu, 180
Malta, 246
Manhattan project, 42, 124, 125, 127, 128, 158, 218, 234

Marcoule, 55, 77, 161, 222
Marshall Islands, 122–3, 128, 133
Martell, E., 240
Marviken, 184–5
Maximum Permissible Dose, 282ff
Mead, M., 146
meltdown, 168–9
merit order, 219–20
Metzger, P., 142, 240
Mexico, 257
Midwest Fuel Recovery Plant, 208, 227
Mihama, 210
Miljoecentrum, 212
Milk Marketing Board, 165
Millstone, 178, 200
Ministry of Agriculture, Fisheries and Food, 108, 165
Minnesota, 151
Mol, 173
Monte Bello Islands: 128
Monticello, 151
Montreal laboratory, 127
Moore, R., 130–31, 218, 220ff
Moss, 144
Mururoa, 154
Mutsu, 210ff

Nader, R., 200
Nagasaki, 32, 120, 121, 124, 135
National Environmental Policy Act, 192–3, 204–5
National Nuclear Corporation, 201
National Radiological Protection Board, 208, 285
National Reactor Testing Station, 63, 76, 112, 167–8
Nautilus, 63, 210
Nepal, 246
Netherlands, 94, 212, 227, 253, 257
Nevada, 132–3, 143, 145, 147, 148, 153, 256
New Jersey Central Power & Light, 178, 223
New Zealand, 153–4
Niagara Mohawk Power, 178
Niederaibach, 209
Nine Mile Point, 178
Nixon, R., 142, 153, 255
Non-Proliferation Treaty, 244ff
Northern States Power, 151
Norway, 128, 244, 257
Novovoronezh, 162
NRU, 161
NRX, 70,
    accident, 158ff
Nuclear Fuel Services, 208
Nuclear Installations Act, 166, 230
Nuclear Installations Inspectorate, 108, 166, 208

Nuclear Materials and Equipment Corporation, 236, 246
Nuclear Power Group, 184
Nuclear Regulatory Commission, 215, 225

Oak Ridge, 47–8, 91ff, 200, 248
Obninsk, 49, 162
Office of Naval Reactors, 176
Oldbury, 54, 183
Open University, 262
Oppenheimer, R., 125
Organization for Economic Cooperation and Development, 59, 174, 229, 259
Orlando, 247
Oskarshamn, 185
Otto Hahn, 210
Oyster Creek, 178, 223

Pacific Gas & Electric, 172, 179
Pakistan, 245, 253, 257
Palisades, 178, 200
Palomares, 237–8
Panama Canal, 155
Partial Test Ban, 132, 147, 153
Pathfinder, 167
Pauling, L., 142
Peach Bottom, 59, 167
pebble-bed reactor, 62
Peru, 153, 154
Pesonen, D., 179
Phénix, 77, 203, 206
Philippines, 257
Pickering, 70, 72, 213–14
Pierrelatte, 91, 94
Pippa, 130
Piqua, 167
Plowshare, 154ff
plutonium production reactors, 47ff
Plym, 128
Poland, 128, 257
Port Hawkesbury, 213
Portugal, 245, 257
Power Reactor Development Corporation, 169, 171, 178
pressurized water reactors (PWRs), 63ff
Price–Anderson Act, 172, 230–31
Priestley, J., 143
Production, Division of Atomic Energy, 127–8
Prototype Fast Reactor, 77, 203, 206, 253
Public Health Service, 90, 142, 145
Pugwash, 140, 247

Quad Cities, 178, 200
Quebec, 94
Queen Elizabeth II, 50

radiation, 25ff
radioactive waste management, 103ff
radioactivity, 25ff
radon, 89, 120
Ramey, J., 170
Rapoport, R., 134, 239, 240
Rasmussen, N., 199
Ravenswood, 180
reactor, 132ff
Reuther, W., 170
Ribicoff, A., 252
Richland, 48
Rickover, H., 176-7
Riley, W., 265
Ringhals, 212
'Rio Blanco', 156-7
Rio Tercero, 255
Robert Emmet Ginna, 178
Rochester Gas & Electric, 178
Rocky Flats fire, 238ff
Roentgen, W., 29, 119
'Rollercoaster', 147
Romania, 257
Rongelap, 133-4
Rongerik, 123, 133-4
Rosenbaum, D., 252
Rotblat, J., 140
'Rulison', 156-7
Russell, B., 139
Russell–Einstein Manifesto, 140

St George, 133, 149
'Salt Vault', 113
San Andreas fault, 179
San Marino, 246
San Onofre, 178, 252
SANE, 141
Santa Barbara spill, 226
Saudi Arabia, 245
Savannah, 210
Savannah River, 128, 213, 238
Schmehausen, 62
Scientists' Institute for Public Information, 146, 200, 205
Seaborg, G., 234
Select Committee on Science and Technology, 200-201
Shell, 215
Shevchenko, 77
Shippingport, 166, 167
Sierra Club, 179, 200
'Simon', 132
Singapore, 257
Sizewell, 183
SL-1 accident, 175-6, 188
Slotin, L., 122
SNAP, 237
Sodium Reactor Experiment, 175
solar energy, 267

South Africa, 245, 253, 257
South of Scotland Electricity Board, 53, 202
Southern California Edison, 178
Spain, 212, 245, 253, 257
Springfields, 127
'Starfish', 147
steam generating heavy water reactors (SGHWRs), 73ff
Sternglass, E., 148ff
Stevenson, A., 141
Stewart, A., 149
Stockholm International Peace Research Institute, 253
Strategic Arms Limitation Talks, 256
Strauss L., 168-9
strontium, 26, 28, 90, 106-7, 111, 133, 141, 145, 146
'Super', 125-6, 128
Super-Phénix, 207
Supreme Court, 153, 178
Swaziland, 246
Sweden, 174, 175, 184, 189, 212, 243, 257
Swiss Association for Atomic Energy, 187
Switzerland, 174, 185ff, 189, 246, 253, 257

Tahiti, 154
tailings, 89, 90
tails assay, 92-3
Taiwan, 257
Tamplin, A., 130-31, 284-5
tank 106T leak, 110-11
Tanzania, 245
Tarapur, 244
Taylor, T., 249ff
Teller, E., 155
Tennessee Valley Authority, 215, 223
Thailand, 257
thermonuclear bomb, 126, 128-9
thorium, 15, 26, 61, 73, 76, 102
Thresher, 177
Thule, 238
thyroid, 107, 134, 142, 148, 164
Togo, 246
Tokai Mura, 175
Torrey Canyon, 226
transport, 98ff
Transuranium Registry, 237, 285
Trawsfynydd, 183
Tricastin, 227
Trico, 46
Trinity, 32
tritium, 104, 126, 156
Trombay, 244, 254
Troy, 133, 138, 149
Truman, H., 128

Tsivoglou, E., 151, 152
Tuohy, T., 163–4
Turkey, 246, 257

Union Carbide, 215
United Auto Workers, 170
United Nations, 123, 129, 139, 173, 243, 244
United Nations Atomic Energy Commission, 173
United Nations Conference on the Human Environment, 154
United Nations Scientific Committee on the Effects of Atomic Radiation, 139
'Upshot-Knothole', 132
uranium, 245
uranium-233, 61, 76, 102
uranium production, 87ff
URENCO, 227
Uterik, 133–4

Vemork, 128
Venezuela, 246
Vermont Yankee, 200
Vietnam, 245, 254
vitrification, 113

WASH-740, 170, 229
WASH-1250, 200
WASH-1400, 202–3, 231, 264
weapons, nuclear, 13, 41, 86, 91, 93, 98, 102, 109, 117, 122–3, 125, 127, 130, 135, 153, 156, 161

Weil, G., 45
West Germany, 59, 62, 90, 94, 114, 174, 209–10, 253, 257
Westinghouse, 183, 196–7, 212, 215, Test Reactor, 175
Whiteshell, 73
Willrich, M., 249ff
Wilson, 201, 256
wind energy, 267
Windscale, 48, 49, 53, 57, 100ff, 107ff, 127, 143, 188
    AGR, 183
    B204 leak, 207
    fire, 162ff
Winfrith, 59, 73
Wright, J., 193
Würgassen incident, 198–9
Wylfa, 54, 61, 183, 222, 223–4

X-ray, 27, 29, 118, 119, 149

Yaizu, 135
Yankee Rowe, 167, 172
Yucca Flats, 132
Yugoslavia, 257

Z-9, 241
Zaire, 46
Zambia, 246
ZEEP, 158
Zinn, W., 168
Zoë, 161